Social and Ecological Interactions in the Galapagos Islands

Series Editors

Stephen J. Walsh, University of North Carolina at Chapel Hill, Chapel Hill, NC, USA

Carlos F. Mena, Universidad San Francisco de Quito, Quito, Ecuador

More information about this series at http://www.springer.com/series/10427

Mary-Ellen Tyler
Editor

Sustainable Energy Mix in Fragile Environments

Frameworks and Perspectives

Editor
Mary-Ellen Tyler
Faculty of Environmental Design
University of Calgary
Calgary, AB, Canada

ISSN 2195-1055   ISSN 2195-1063 (electronic)
Social and Ecological Interactions in the Galapagos Islands
ISBN 978-3-319-69397-2   ISBN 978-3-319-69399-6 (eBook)
https://doi.org/10.1007/978-3-319-69399-6

Library of Congress Control Number: 2017962297

© Springer International Publishing AG 2018
This work is subject to copyright. All rights are reserved by the Publisher, whether the whole or part of the material is concerned, specifically the rights of translation, reprinting, reuse of illustrations, recitation, broadcasting, reproduction on microfilms or in any other physical way, and transmission or information storage and retrieval, electronic adaptation, computer software, or by similar or dissimilar methodology now known or hereafter developed.
The use of general descriptive names, registered names, trademarks, service marks, etc. in this publication does not imply, even in the absence of a specific statement, that such names are exempt from the relevant protective laws and regulations and therefore free for general use.
The publisher, the authors and the editors are safe to assume that the advice and information in this book are believed to be true and accurate at the date of publication. Neither the publisher nor the authors or the editors give a warranty, express or implied, with respect to the material contained herein or for any errors or omissions that may have been made. The publisher remains neutral with regard to jurisdictional claims in published maps and institutional affiliations.

Printed on acid-free paper

This Springer imprint is published by Springer Nature
The registered company is Springer International Publishing AG
The registered company address is: Gewerbestrasse 11, 6330 Cham, Switzerland

In Memory of Dr. Julie Rowney

Co-chair of the 2014 World Summit in the Galapagos, Professor of Human Resources and Organizational Dynamics, Haskayne School of Business, University of Calgary. Director of the Centre for International Management, Director of the OLADE Project for Latin American Energy, Director of the Master of Science in Sustainable Energy Development degree program delivered jointly with the Universidad San Francisco de Quito

"If I have seen further, it is by standing on the shoulders of giants"-Isaac Newton, 1675

Foreword

Energy is generally thought of as an important economic commodity. However, energy is much more than just an economic commodity. It is a fundamental life-support system just like air, water, and food. The connection between human welfare and energy makes it a strategic geopolitical resource. Because energy is so fundamental to human development, it has economic value. But just as importantly, it has social and ecological value. Ecological systems and social systems are both energy-dependent systems. Energy decision-making cannot be isolated from its social and ecological context. Sustainable energy development means viewing energy as a critical and integral part of social and ecological systems.

Sustainable energy mix means understanding energy sources in their social-ecological and geographic contexts. The term "social" is used here to include cultural norms, traditions, values, religious beliefs, indigenous worldviews, economic, institutional, political frameworks, demographics, and technology use. The term "ecological" is used to include biological, geological, and climatic conditions, as well as terrestrial ecosystem dynamics, aquatic or marine ecosystem dynamics, landscape ecology, and plant and animal species and populations. Fragile environments include geographically isolated areas such as islands and remote areas with extreme climatic conditions. This term also incorporates protected areas designated for important cultural and heritage value and biodiversity. This includes national parks and designated World Heritage sites. The Galapagos Islands represent a microcosm of fragile environments all over the world that are experiencing increasing energy demands from marine and land-based tourism, local community population growth, resource development projects, and increasing demand for goods and services. The challenge facing fragile environments is the risk of "trading off" ecological and cultural heritage protection for increasing flows of economic goods and services. Sustainable energy mix development in fragile environments offers an alternative approach to this trade off dilemma. Effective sustainable energy mix planning in the Galapagos and other fragile environments requires moving away from conventional single-minded technical and economic thinking. The focus of sustainable energy mix planning is finding social, ecological, technological, and economic interrelationships that work in specific landscape and marine contexts.

The methodological challenge is how to "frame" complex human-ecological interconnections operating across multiple spatial and temporal scales. A sustainable approach to energy development focuses on functional and structural social-ecological-economic interconnections in specific contexts and at specific scales. This approach is necessary to identify and support an energy mix strategy capable of linking long-term ecosystem behavior with human activity systems. As such, it offers an alternative to more conventional institutional thinking in which social, ecological, and economic interests are prioritized and traded off against each other.

In a "real-world" context of sustainability practice, practitioners and theorists need to understand the importance of social-ecological systems as well as the methodological and institutional options for their management. Stakeholder engagement is a primary mechanism through which contextual answers to sustainable energy mix design can emerge. This is an integrative, interdisciplinary, and transdisciplinary process involving practitioners, researchers, and stakeholders in the creation of new knowledge and customized knowledge. This process of co-discovery is an important social learning process characteristic of a sustainable energy mix in fragile environments approach.

Too often, assumptions are made about problems and solutions with insufficient information or knowledge about the context in which such problems exist. The 2014 World Summit focused on framing sustainable energy mix in a way that would enable researchers and practitioners to have a common conversation about the contextual issues, the driving forces, and what we know as well as what we don't know. The collection of papers in this volume address the "diagnose-design-do-develop" research framework used to organize the 2014 World Summit as illustrated below. Collectively, this volume represents a collaborative overview of the multiple dimensions of sustainable energy mix in fragile environments. This collection provides a curated body of research and practice experiences and future directions for understanding the challenges and best practices involved in planning, designing, and managing sustainable energy mix solutions for fragile environments.

Foreword

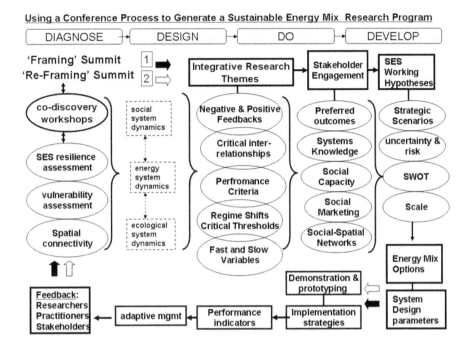

I have had the privilege of working with colleagues at both the University of Calgary and the Universidad San Francisco de Quito for several years in the delivery of a Master of Science in Sustainable Energy Development in both Calgary and Ecuador. This has enabled me to see firsthand the parallels between development pressures on fragile environments in Canada's northern coastal marine zone and the Galapagos. I believe the work represented in this book is important. The loss of the cultural and biological diversity represented by fragile environments both north and south is a loss to humanity and our common future. There is no substitution, and once these places and cultures are gone, they are gone forever. We owe it to future generations to show that we can work across disciplinary and geopolitical lines to find sustainable social-ecological energy solutions.

University of Calgary, Calgary, AB, Canada   Mary-Ellen Tyler

Series Foreword

In this latest contribution to Springer's Galapagos Book Series, "*Social and Ecological Interactions in the Galapagos Islands,*" Stephen J. Walsh and Carlos F. Mena, *Editors*, Mary-Ellen Tyler, Guest Editor and Professor, Faculty of Environmental Design, University of Calgary, Canada, has developed, with her colleagues, an interesting and timely book that examines several topics associated with sustainable energy mix in fragile environments. Reported through a diverse set of chapters written by a highly qualified group of authors, they invoke multiple frameworks and perspectives to examine the challenges related to energy supply, energy consumption, and energy policies that are highly germane to islands and other fragile settings. The chapters also describe the impacts of energy mix on economic development, environment, and communities, with implications for resident and migrant populations who come to islands and other similarly fragile places, increasingly, as tourists and/or as workers to support the burgeoning tourism industry.

The vision for the book emanated from a workshop held at the Galapagos Science Center on San Cristobal Island, Galapagos Archipelago of Ecuador, a facility dedicated to island research, education, and community outreach and achieved through a collaborative partnership between the University of North Carolina at Chapel Hill, USA, and the Universidad San Francisco de Quito, Ecuador. With a focus on island ecosystems and threats to their sustainability, the Galapagos Science Center hosted the workshop participants and encouraged the ideas they expressed on energy mix, ecological and social fragility, and island sustainability. With tours of the local communities, assessment of local energy services, characterization of transport and offloading facilities, understanding household demands for consumer products, and the use of power generation technologies, the workshop participants observed how and why the Galapagos Islands are considered a "natural laboratory," but not only relative to understanding the environment, but, so too, for understanding the social factors that drive the expanding human dimension and the challenges to sustainability of fragile and protected environments. With the Galapagos National Park responsible for the management and protection of 97% of the land area of the archipelago as well as the second largest marine reserve in the world, the world renowned status of the Galapagos Islands heightens the importance of energy mix in the islands and the potential for land and marine degradation through direct and indirect consequences of the expanding human dimension related to energy and the generation and consumption of electrical power in fragile settings.

In the Galapagos Islands, long identified by the iconic species that live there, most notably, giant tortoises, marine iguanas, and the Darwin finches, most of the power consumed in the islands is provided through the use of generators that run on diesel fuel imported from the Ecuadorian mainland, nearly 1000 km away. Transported to the archipelago on cargo and fuel ships and linked to the expanding consumptive demands of residents and tourists, an ever-increasing risk to the environment is generated as a consequence of the number and size of supply ships that are needed to support the needs of hotels, restaurants, residences, and commercial enterprises with electricity as well as fuel for imported trucks, taxis, boats, and cars that are part of the increasing human dimension throughout the four populated islands. On San Cristobal Island, for instance, three large wind turbines contribute to the power grid that supports the communities of El Progresso, located in the highlands, and the coastal community of Puerto Baquerizo Moreno as well as the island's airport that accommodates flights from the continent as well as flights between the islands. Cargo and cruise ships, pleasure crafts, and smaller, commercial boats ferry people and goods between islands using fuel imported from the mainland, but this occurs not without their problems. Over several years, ships have grounded and spills have occurred that highlight the concerns for energy transport and distribution, best exemplified through the 2001 sinking of the *Jessica* and the ecological problems that it generated throughout the Galapagos Marine Reserve. In addition to the use of petroleum-based products in the Galapagos, photovoltaic solar panels are used to locally generate a small amount of power at selected government and nongovernment facilities, including the Galapagos Science Center; however, diesel generation of power remains the primary way of supporting the electrical grid on San Cristobal Island and throughout the populated islands of the Galapagos archipelago.

In short, this book addresses vital concerns germane to social and ecological well-being of humans as well as plants and animals, both terrestrial and marine, throughout the Galapagos Islands and in other geographic settings where fragile and sensitive ecosystems occur. This book provides an important and supplementary dimension to the Galapagos Book Series that highlights social-ecological pressures on the islands as well as other fragile places, and offers insights into the sustainability of energy mix in areas of increased visitation by migrant populations, including national and international tourists and residents linked to both economic development and resource conservation, in the case of the Galapagos, that occur in and around protected areas.

University of North Carolina at Chapel Hill, Chapel Hill, NC, USA	Stephen J. Walsh
Universidad San Francisco de Quito, Quito, Ecuador	Carlos F. Mena

Preface

Energy is a critical life-support system in both social systems and ecological systems. In remote and fragile environments such as islands and remote locations with extreme climate conditions, access to sustainable energy is essential for the welfare of local communities. Many fragile environments are experiencing increasing resource development and tourism pressures. Appropriate energy sources and technologies are critical for maintaining environmental quality and quality of life in the face of increasing energy demand. However, sustainable development issues arise in geographic locations lacking or with limited access to conventional energy sources. There is a critical link between affordable clean energy and sustainable development. In remote and fragile environments, situational access to conventional and unconventional energy sources offers an opportunity to create a sustainable mix of energy sources and technologies customized to fit local social, cultural, environmental, and economic circumstances.

As a designated United Nations Organization for Education, Science and Culture (UNESCO) World Heritage Site, Marine Reserve and National Park, the Galapagos Islands of Ecuador are one example of such a fragile environment found in a remote offshore location with local communities and marine and terrestrial ecosystems of historical and international significance. The need to manage growing land- and water-based international eco-tourism pressures while striving to improve the quality of life for local communities and conserve the ancient and biologically significant flora and fauna of the islands makes the Galapagos a microcosm of the sustainable energy mix challenges facing remote and fragile cultural and biophysical environments worldwide.

The University of Calgary, Alberta, Canada, and the Universidad San Francisco de Quito, Ecuador, have worked together with the Canadian International Development Agency (CIDA) and the Organizacion Latinoamericana de Energia (OLADE) to successfully deliver a Master of Science in Sustainable Energy Development degree program over a number of years. Given this partnership experience and the growing awareness of the importance of energy mix in achieving sustainable socioeconomic development, representatives of the Sustainable Energy Development program from both Calgary and Quito undertook the planning for a

World Summit on the island of San Cristobal in the Galapagos to address the importance of sustainable energy mix in fragile environments. The Inter-American Development Bank (IDB) Energy Innovation Center and Mount Royal University's Institute for Environmental Sustainability assisted in organizing this international gathering of experts and providing financial sponsorship. Summit co-chairs, Dr. Diego Quiroga of the Universidad San Francisco de Quito and the late Dr. Julie Rowney of the University of Calgary, worked with the Summit Program Planning Committee (Dr. Anil Mehrotra, Dr. Mary-Ellen Tyler, Allan Ingelson, and Dr. Irene Herremans of the University of Calgary, Dr. Michael Quinn of Mount Royal University, and Annette Hester of IDB) for over 2 years to make the Summit happen.

Over 50 policymakers, business leaders, energy and economic development practitioners, and researchers in social, environmental, and technological dimensions of energy and sustainability were invited to participate in a working Summit to examine the driving forces and critical factors affecting energy mix and socioeconomic development in the context of environmental sustainability. Summit objectives were to:

- Convene a group of international experts and practitioners with interdisciplinary backgrounds and experience related to energy mix issues in fragile social-ecological environments.
- Identify relevant theories and best practice contributions to the Summit's core themes.
- Identify research priorities for designing and customizing energy mix in different biogeoclimatic and cultural contexts.
- Use the Galapagos Islands venue as a "case study" of energy mix issues and possible solutions.
- Create a network of international practitioners and researchers to carry forward Summit results for testing in a wide variety of locations and sectors.

The Summit was held in San Cristobal in the Galapagos from July 20 to 24, 2014. Small group workshops were organized around five key case study presentations which explored issues, knowledge and practice gaps, and possible cross-sector approaches to dealing with sustainable energy mix in fragile environments. Five themes emerged from the Summit workshops and provide direction for further research and practice exploration:

- Need for new institutional frameworks
- Fossil fuel subsidy dependency
- Importance of social capacity
- Potential capacity of renewables
- Receptivity to technology transfer

This volume presents 11 invited papers that address different aspects of the Summit's 5 themes and sustainable energy mix design in the Galapagos and comparative contexts. The first paper provides a historical overview of the Galapagos Islands with a focus on the evolving biocomplexity of the islands' social and ecological systems. Some significant events are identified that have affected system feedbacks and interactions over time in the Galapagos as the islands have moved from a relative state of isolation to daily flights of international tourists. Expanding fishing industry pressures and growing social, economic, and technological interconnections with the mainland present a complex context for developing a sustainable energy mix. The second paper explores the driving forces affecting the need for sustainable energy mix solutions, and the third paper provides two project examples illustrating the importance in practice of social engagement and education. The fourth paper examines the potential for biofuels and the use of biofuels in the Galapagos, while the fifth paper looks at the legal and institutional issues related to renewable energy mix development. The sixth paper illustrates the importance of life cycle analysis in energy mix planning, and the seventh paper reviews renewable energy development experiences and lessons learned in Ecuador's Amazon region. Papers eight and nine document two different approaches to waste management in fragile environments. The tenth paper examines fragile cultural and marine environments in a northern latitude context with parallels to the Galapagos. The eleventh and final paper lays out an approach to energy mix planning for fragile environments that can be adapted to different locations and institutional contexts. Collectively, these chapters provide a comprehensive framework for understanding the multidisciplinary dimensions of sustainable energy mix. Much of the content of the papers included in this volume is based on professional practice and grounds the discussion in the operational reality of what has and has not worked.

As guest editor, I first want to acknowledge the importance of the support and assistance received from the series editors, Stephen Walsh and Carlos Mena, in ensuring the completion of this work. I also want to acknowledge the tremendous cooperation and commitment from the authors of the papers represented in this collection. Irene Herremans deserves special thanks for her assistance in helping me at critical times. I want to recognize the importance of the participants in the 2014 World Summit in the Galapagos. The contents of this volume represent their experience, wisdom, expertise, and professional and personal commitment to a sustainable energy and their contributions to Summit workshop discussions.

Finally, it may initially seem odd that a book focusing on energy would be included in a series focusing on social and ecological interaction in the Galapagos Islands. It is my hope, as guest editor, that the contents of this volume will convince readers of the impact that energy mix decisions have on the well-being of future social and ecological interactions in the Galapagos and in fragile environments all over the world.

University of Calgary, Calgary, AB, Canada　　　　　　　　　　Mary-Ellen Tyler

Series Preface

Galapagos Book Series, "Social and Ecological Sustainability in the Galapagos Islands"

When we developed the Galapagos Book Series and selected the initial book topics to launch the series, we hoped that guest editors and authors would cooperate to represent important and fascinating elements of the Galapagos Islands early in the series. *Science and Conservation in the Galapagos Islands: Frameworks & Perspectives*, Stephen J. Walsh & Carlos F. Mena, editors (2013), advocates an interdisciplinary perspective for addressing many of the most compelling challenges facing the Galapagos Islands that extend across the social, terrestrial, and marine subsystems. *Evolution from the Galapagos: Two Centuries after Darwin*, Gabriel Trueba & Carlos Montufar, editors (2013), advances our understanding of evolution, a key element of life and adaptation in the Galapagos Islands. *The Galapagos Marine Reserve: A Dynamic Social–Ecological System*, Judith Denkinger & Luis Vinueza, editors (2014), addresses the nature of the coupled human–natural system in the Galapagos Islands and describes some of the key factors that affect social and ecological vulnerability, dynamics, and island sustainability. *Darwin, Darwinism and Conservation in the Galapagos Islands*, Diego Quiroga & Ana Sevilla, editors (2016), examines the meaning and essence of Darwin and Darwinism in the Galapagos and beyond. His ideas shook the world of science and continue to give meaning and explanations of life and the adaptive capacity of species in the Galapagos and around the globe. *Disease Ecology of Galapagos Birds*, Patricia Parker, editor (2017/2018), addresses the central elements of birds in the Galapagos Islands associated with colonization, pathogens, hosts and parasites, and evolution. And now, *Sustainable Energy Mix in Fragile Environments: Frameworks and Perspectives*, Mary-Ellen Tyler, editor (2018), examines sustainable energy mix economic development, communities, and fragile and sensitive environment. *Understanding Invasive Species in the Galapagos Islands: From the Molecular to the Landscape*, Maria de Lourdes Torres & Carlos Mena, editors (2018), examines the introduction of alien species into the Galapagos Islands and the multiscale assessment of them through, for instance, DNA approaches as well as satellite remote sensing to understand their establishment, ecology, spread, and eradiation.

It was not until Charles Darwin's famous visit in 1835—which helped inspire the theory of evolution by natural selection—that the Galapagos Archipelago began to receive international recognition. In 1959, the Galapagos National Park was formed, and in 1973, the archipelago was incorporated as the 22nd province of Ecuador. UNESCO designated the Galapagos as a World Heritage Site in 1978, a designation to honor the "magnificent and unique" natural features of the Galapagos and to ensure their conservation for future generations. These islands were further deemed a Biosphere Reserve in 1987, and the Galapagos Marine Reserve was created in 2001. The Marine Reserve was formed as a consequence of the 1998 passage of the Special Law for Galapagos by the Ecuadorian government that was designed to "protect and conserve the marine and terrestrial resources of the Islands." Development of the tourism industry has more than tripled the local population in the past 15 years, thereby exerting considerable pressure on the Galapagos National Park and the Marine Reserve. The residential population has grown from approximately 10,000 in 1990 to nearly 30,000 residents today, and national and international tourism has increased from approximately 40,000 visitors in 1990 to now in excess of 225,000. The impacts of the human dimension in the islands have been both direct and indirect, with consequences for the social, terrestrial, and marine subsystems in the Galapagos Islands and their linked effects. Further, the historical exploitation of lobster and sea cucumber, globalization of marine products to a national and international market, and the challenges imposed by industrial fishing outside of the Reserve and illegal fishing and shark fining outside and inside the Reserve combine to impact the social and ecological vulnerability of the Galapagos Marine Reserve in fundamental ways. In addition, exogenous shocks, such as ENSO events as a disturbance regime on Galapagos corals and marine populations, national and international policies and institutions on regulation and management, and the "pushes" and "pulls" of economic development and population migration, including tourism, shape and reshape the Galapagos Islands—its resources, environments, people, and trajectories of change.

Globalization is a process that affects island ecosystems and poses social-ecological threats to their sustainability. This book explores important energy-related topics with implications for the Galapagos and other fragile and sensitive island ecosystems around the globe. It is a wonderful addition to the series and another topic that resonates within and outside of the Galapagos Archipelago of Ecuador.

University of North Carolina at Chapel Hill, Chapel Hill, NC, USA	Stephen J. Walsh
Universidad San Francisco de Quito, Quito, Ecuador	Carlos F. Mena

Contents

1 **Galapagos: A Microcosm of Sustainable Energy Mix in Fragile Environments**... 1
Diego Quiroga

2 **Driving Forces and Barriers for a Sustainable Energy Mix in Fragile Environments: North–South Perspectives**.............. 21
Michael S. Quinn

3 **Climate Change Policy as a Catalyst for Sustainable Energy Practice: Examples from Mainland Ecuador and the Galapagos**... 33
Irene M. Herremans and Mary-Ellen Tyler

4 **Biofuels in the Energy Mix of the Galapagos Islands**............. 49
Irene M. Herremans and Arturo Mariño Echegaray

5 **Policies and Laws and Island Environments**.................... 57
Allan Ingelson and Christopher Phillip

6 **Using Life Cycle Assessment to Facilitate Energy Mix Planning in the Galapagos Islands**................................. 93
Eduard Cubi, Joule Bergerson, and Anil Mehrotra

7 **Sustainability of Renewable Energy Projects in the Amazonian Region**.................................. 107
Juan Leonardo Espinoza, José Jara-Alvear, and Luis Urdiales Flores

8 **Estimation of Landfill Methane Generation from Solid Waste Management Options in the Galapagos Islands**................. 141
Rodny Peñafiel, Lucila Pesántez, and Valeria Ochoa-Herrera

9 **Biodigesters as a Community-Based Sustainable Energy Solution**... 153
Elizabeth Romo-Rábago, Irene M. Herremans, and Patrick Hettiaratchi

10 **Sustainable Energy Mix + Fragile Environments in Canada's Northern Coastal Zone: Is Technology Enough?**................. 163
Mary-Ellen Tyler and Allan Ingelson

11 **Sustainable Energy Mix in Fragile Environments: A Transdisciplinary Framework for Action**..................... 183
Mary-Ellen Tyler and Irene M. Herremans

Index.. 195

Contributors

Joule Bergerson Energy and Environment Systems Group (EESG), Centre for Environmental Engineering Research and Education (CEERE), Department of Chemical and Petroleum Engineering, Schulich School of Engineering, University of Calgary, Calgary, AB, Canada

Eduard Cubi Energy and Environment Systems Group (EESG), Centre for Environmental Engineering Research and Education (CEERE), Department of Chemical and Petroleum Engineering, Schulich School of Engineering, University of Calgary, Calgary, AB, Canada

Arturo Mariño Echegaray Faculty of Environmental Design PhD Program, University of Calgary, Calgary, AB, Canada

Juan Leonardo Espinoza Department of Electrical and Electronics Engineering, University of Cuenca, Cuenca, Ecuador

Luis Urdiales Flores Empresa Eléctrica Regional Centrosur C.A., Cuenca, Ecuador

Irene M. Herremans Haskayne School of Business, University of Calgary, Calgary, AB, Canada

Patrick Hettiaratchi Department of Civil Engineering, Schulich School of Engineering, University of Calgary, Calgary, AB, Canada

Allan Ingelson Faculty of Law, University of Calgary, Canadian Institute of Resources Law, Calgary, AB, Canada

Faculty of Law, University of Calgary, Calgary, AB, Canada

José Jara-Alvear Center for Development Research (ZEF), University of Bonn, Bonn, Germany

Anil Mehrotra Energy and Environment Systems Group (EESG), Centre for Environmental Engineering Research and Education (CEERE), Department of Chemical and Petroleum Engineering, Schulich School of Engineering, University of Calgary, Calgary, AB, Canada

Valeria Ochoa-Herrera Colegio de Ciencias e Ingenierías, Universidad San Francisco de Quito, Quito, Ecuador

Rodny Peñafiel Colegio de Ciencias e Ingenierías, Universidad San Francisco de Quito, Quito, Ecuador

Lucila Pesántez Colegio de Ciencias e Ingenierías, Universidad San Francisco de Quito, Quito, Ecuador

Christopher Phillip Student-at-law, British Columbia (B.C.) Hydro, Vancouver, BC, Canada

Michael S. Quinn Mount Royal University, Calgary, AB, Canada

Diego Quiroga College of Biological and Environmental Sciences, Universidad San Francisco de Quito, Quito, Ecuador

Elizabeth Romo-Rábago The Social License Consortium, Calgary, AB, Canada

Mary-Ellen Tyler Faculty of Environmental Design, University of Calgary, Calgary, AB, Canada

Chapter 1
Galapagos: A Microcosm of Sustainable Energy Mix in Fragile Environments

Diego Quiroga

Introduction

The Galapagos Islands are near the equator in the Pacific Ocean approximately 1000 km from the closest land area, the mainland of Ecuador. This archipelago houses a high number of endemic species unique in the world. With 95% of the endemic animals and plants still present, the Galapagos is considered one of the best conserved archipelagoes of the world. Diverse ocean currents, some of which, like the Cromwell Subequatorial Current and Humboldt or Peruvian Current, are cold and nutrient rich and are associated with upwelling cells. Others like the Panama Current, carry warm waters. These currents produce contrasting conditions of temperature and precipitation and generate a unique set of conditions that allow for the presence of the characteristic flora and fauna in a relatively small area. Many of the islands have tall volcanoes reaching more than 1000 m above sea level trapping the clouds and moisture; the different altitudinal levels that produce diverse ecological zones constitute an opportunity for divergent adaptations.

Isolation is a key aspect for the Galapagos as a biological system, and it is a major factor in the evolution and diversification of its different species. In order for organisms to arrive naturally to the Galapagos, they must be able to withstand the long oceanic voyage. Some of these organisms came floating on their own or on natural drafts; others came carried by the wind and air currents or carried by other organisms, usually birds. On occasions, some of these plants and animals could not only endure the trip to the Galapagos but were also able to establish themselves on the new environment, reproduce, and evolve in unique and surprising ways. Once they arrived to the Galapagos, the organisms colonized different islands, and the

D. Quiroga (✉)
College of Biological and Environmental Sciences, Universidad San Francisco de Quito, Quito, Ecuador
e-mail: dquiroga@usfq.edu.ec

© Springer International Publishing AG 2018
M.-E. Tyler (ed.), *Sustainable Energy Mix in Fragile Environments*, Social and Ecological Interactions in the Galapagos Islands, https://doi.org/10.1007/978-3-319-69399-6_1

relative isolation between islands produced further processes of speciation and rapid island evolution. These biogeographical conditions resulted in the creation of new species from a few successful individuals that arrived to the islands. Many endemic species evolved as a result of processes of speciation and diversification and adaptive radiation (Valle 2013). In the case of the Galapagos, some classic examples of these processes of adaptive radiation include several species of plants including *Scalesia*, invertebrates such as the land snails, and many species of animals such as the Darwin finches, the mockingbirds, lizards, iguanas, geckos, and the famous Galapagos tortoises. As in other oceanic islands, there is a disharmonic biota in the Galapagos as there are some species and groups of organisms that are able to arrive, reproduce, and adapt to these new conditions, whereas other organisms cannot do so. One of the most important consequences of this process of evolution is the low numbers of individuals of each species and their specific and limited distribution range. Such is the case for the Galapagos flightless cormorant, the Galapagos penguin, the mangrove finch, and many species of plants. These characteristics make these species very vulnerable to natural or human-caused changes such as El Niño events or oil spills. These vulnerable and iconic animals are important to a growing tourism industry.

Organisms and ecosystems in oceanic islands are especially vulnerable to invasive species; most extinctions of birds and reptiles have occurred in island ecosystems where animals and plants have not developed the capacity to defend themselves from exotic introductions (Simberloff 2013). In the Galapagos there are some animal populations that have very low numbers including several species of plants, the mangrove finch, the Floreana mockingbird, some species of tortoises, rice rats, the Galapagos penguins, flightless cormorants, and the pink land iguana. Cheap hydrocarbons have played an important role in disturbing the isolation of the islands. When new ships and vessels started using petroleum derivatives to move people and cargo around the world, the Galapagos become accessible to yachts carrying tourists, scientists, and cargo boats bringing goods from distant lands. When coal and petroleum powered ships entered the scene, people settled permanently in the Galapagos almost 200 years ago. Even before engines were used to move the large boats, sails allowed large numbers of vessels carrying pirates and whalers to overhunt tortoises, whales, and sea lions, dangerously lowering the populations of these iconic species. They also brought with them some of the early introduced plants and animals. Thus when Darwin arrived in the Beagle, he mentions 17 introduced species. However, it was when ships powered by hydrocarbons and later airplanes started to increase the connectivity between the mainland and the Galapagos that the amount of introduced species grew at an exponential rate, and these introductions became increasingly responsible for most of the changes in the ecosystems and the threats to the endemic species. Availability of cheap fossil fuels has made it possible for humans to arrive in greater numbers and to live with more comfort in the islands, but also eroded the needed isolation that protects the native and endemic species from diseases, competition, and predation. Technological changes and increased energy availability have also meant an increase in the number of tourists and the expansion of extractive industries like fishing. Cheap hydrocarbons have generated a positive feedback cycle as greater degrees of connectivity also increase the amount of people living in the Galapagos and their energy needs which are largely supplied by fossil fuels.

Endemism and Vulnerability

A high degree of endemics can be seen in the case of reptiles, birds, insects, fishes and algae, other terrestrial invertebrates, and vascular plants (Peck 1996; Tye 2001; Valle 2013). Most of the charismatic species of the Galapagos, such as the two species of sea lions (*Zalophus wollebaeki* and *Arctophoca galapagoensis*), the Galapagos penguin (*Spheniscus mendiculus*) and the flightless cormorant (*Phalacrocorax harrisi*), the land (*Conolophus subcristatus*) and marine iguana (*Amblyrhynchus cristatus*), and the 11 extant species of tortoises are endemic species. Many of these species have undergone the important process of island adaptation. Similarly land snails have undergone a spectacular process of speciation with 71 species (Valle 2013); the Darwin's finches have some 14 species living in the Galapagos, depending on the classification system used; and the giant tortoises that were originally 15 species have 10 still present in the Islands. Among plants, the genus *Scalesia* (*Asteraceae*) with 15 species and 19 taxa including subspecies and varieties is an incredible example of radiation, *Alternanthera* (*Amaranthaceae*) with 14 species and 20 taxa, and *Opuntia* (*Cactaceae*) has 6 species and 14 varieties (Valle 2013). The distance between the islands explains in part the process of speciation that these Galapagos species have experienced. The isolation of the different species within the archipelago is also threatened by increased connectivity between the islands as residents and visitors are increasingly traveling from one island to the other (Ouvrard and Grenier 2010).

The arrival of introduced species that are predators or disease vectors, or that compete for habitat or food, is a major concern in the many oceanic islands. The list of introduced plants in the Galapagos is expanding at a fast rate, and they have become the major biological threat to the Islands. At the moment, there are almost 900 species of introduced plants more than native ones found in 46 Islands. Most of them (560) were introduced for agricultural purposes, and 94 were accidental introductions; only four species have been eradicated. Most introduced species have arrived in the last 30–50 years (Buddenhagen, 2006; Gardener et al. 2013; Guézou et al. 2016). It has been calculated that 332 of these plant species have been naturalized, and 32 of them are invasive (Jaramillo and Guezou 2012; Tye 2001, 2008). In 2006 3 species were found to be extinct, and 20 were critically endangered. In the case of vertebrates, 14 species are considered extinct. The main causes of extinction include habitat loss and or fragmentation and arrival of introduced species including agents of infection, hunting, and climate change (Jiménez-Uzcátegui et al. 2008). There are three species of reptiles and two of birds that are critically endangered: the mangrove finch (*Camarhynchus heliobates*) and the Floreana mockingbird, *Nesomimus trifasciatus*, which are threatened by the introduction of insects like ants, wasps and flies, and bird species and mammals like cats and dogs (Jiménez-Uzcátegui et al. 2008; Valle 2013). Two species of birds found in the western islands of the Galapagos, the penguin (*Spheniscus mendiculus*) and the flightless cormorant (*Phalacrocorax harrisi*), are considered endangered by UICN (Jiménez-Uzcátegui et al. 2008). By 2014, the number of introduced invertebrates had reached 762 taxa (Causton et al. 2006). These invasive species include fire ants (*Wasmannia auropunctata* and *Solenopsis geminata*), wasps (*Polistes v.* and *versicolor*), cottony cushion scale (*Icerya purchasi*), and a bird ectoparasite (*Philornis downsi*). The introduction of infection agents, vectors, and hosts poses a major risk factor that could lead to extinction of species. Some diseases such as avian malaria, West Nile Virus, and diseases transmitted from dogs to sea lions such as brucellosis and rabies constitute important risks to the local fauna. As the number of boats carrying cargo and airplanes carrying people and products has increased in the last few years, the arrival of introduced organisms has become a major threat.

The Galapagos as a Frontier for Extractive Economies

A desire for more connectivity with the mainland is the result of a cultural framework that is shared by many of the inhabitants of the Galapagos. This cultural framework sees the Galapagos as a frontier and as a resource-rich area for the expansion of different extractive economies (Quiroga 2009). The local inhabitants of San Cristobal, for example, are proud to tell how they built their own airport in the 1980s so they could have directs flights to the continent. For them, being able to have more and easier possibilities for contact with the mainland is essential to maintain their standard of living. This means access to what many local residents consider to be adequate education, health, and goods that are not found in the Islands. Increased connectivity of the Galapagos and the mainland has been part of the hopes and expectations of the local inhabitants since the time of the early colonists. The extraction of natural resources was one of the main drivers of this increased interaction between the mainland and the islands. The extraction of resources started early, with tortoises being taken away by privateers and whalers because of the capacity of these animals to endure long periods of time without any food or water. This idea of the Galapagos as a source of goods to be exploited and sometimes extracted continued during the time of colonization by Ecuadorians that started in the 1830s (González et al. 2008). In the nineteenth century, a series of colonization attempts were organized by the Ecuadorian government. These early attempts resulted in the creation of different failed settlements on the island of Floreana. In 1870, a colony was established in San Cristobal by Manuel J. Cobos, who failed in his attempt to extract orchilla, a lichen used as a dye in the textile industry. He later created a successful cattle farm and a sugar and coffee plantation (Latorre 2001). As in the previous case, this colony was also based to a large extent on outlaws and prisoners, some of whom eventually killed Cobos, who they accused of being a brutal oppressor (Latorre 2001). A similar pattern occurred on Isabela where cattle farms and plantations were also created before the end of the nineteenth century. Few boats, usually owned by the owners of the plantations arrived sporadically to the Islands. Once these haciendas were dissolved, the people who stayed divided the land into small farms or *fincas*. Until the 1960s and 1970s, most Ecuadorians viewed the Galapagos as a frontier—a remote and harsh place, where the land could be tamed through hard labor and the creation of agriculture and cattle farms.

In the twentieth century, the extraction of marine resources from the Galapagos included grouper, lobsters, and sea cucumbers. Colonists living on their farms in the highlands descended to the coastal areas, initially only during certain times of the year, to participate in different fishing activities. Large boats came to the islands from countries like Japan to purchase fish, lobsters, turtles, and other products from the local inhabitants. Local people started also to fish for grouper (*Mycteroperca olfax*), which they salted and dried and sent to the mainland. Green and red spiny lobsters (*Panulirus penicillatus* and *P. gracilis*) which were harvested in the 1960s became major exports in the 1980s (Hearn 2008). In the second part of the twentieth century, technological changes such as diesel engines for the larger wooden fishing boats and a fleet of smaller fiberglass boats (known locally as fibras and pangas) outfitted with outboard gasoline engines between 40 and 75 HP were introduced in the Galapagos allowing fishing in distant islands and sea mounds. A system of diving called *hookah* based on small gasoline-powered air compressors that allowed them to dive deeper to fish for lobster and sea cucumbers was also introduced. Power generators brought to the island allowed for refrigeration that fishers needed to keep their product fresh. These technological changes resulted in an increase in the capacity of fishers to capture and overexploit a variety of marine organisms including lobsters, sea cucumbers, and sharks. These improvements were driven by growing imports of subsidized hydrocarbons to the islands that fueled the boom of these fisheries and later their collapse. As Southeast Asian economies improved in the 1980s and 1990s, there was increased demand for sea cucumbers (*Isostichopus fuscus*), and many residents and newly arrived migrants became involved in the fishing of these echinoderms. The Galapagos National Park (GNP) tried to control sea cucumber overfishing in the middle of the 1990s which resulted in tensions and conflicts (Ben-Yami 2001; Hearn 2008). Between 1995 and 2005, several strikes and protests organized by the fishers and the Galapagos conservation and tourism sectors paralyzed the GNP, creating instability and mismanagement (Hearn 2008; Quiroga 2009). In the last decade, the economic importance of the fisheries has diminished in a significant way. Whereas in 2003, fishing was a major part of the economy of the islands, by 2006 fishing made up less than 4% of the general income (Watkins and Cruz 2007). Of the previously 1200 registered fishers, there are only some 300 active ones in the Galapagos at the moment.

In the last 10 years, many newcomers and local residents including fishers, have shifted to new economic areas and have found jobs in the tourism sector. Thus fishers are now employed as sailors, captains, dive masters, and some manage their own sport fishing and day tour operations (Engie 2014). Regulating new activities such as "pesca vivencial" (artisanal fishing with tourists) have now become major concerns for some Park planners and managers (Schuhbauer and Koch 2013; Engie and

Quiroga 2013). New tourism permits have been given to fishers allowing them to visit different areas and to profit from the growing tourism industry. A large part of the local economy now is in one way or another based in tourism, which accounts for more than 60% of the economy (Kerr et al. 2004). The Galapagos National Park in an attempt to ensure a safe and comfortable experience for the tourists has forced fishers and local people to buy larger four-stroke engines and bigger boats. Many fishers and local people who are now engaged in tourism have acquired larger speedboats and have traded their 40 and 75 HP engines for 150 or 200 HP engines (Ouvrard and Grenier 2010).

Cruise Boats and Tourism

In the middle of the twentieth century, many scientists who were concerned with the long-term sustainability of the Islands thought that tourism could become a way of preventing the destruction of natural resources and providing money for conservation. In a report from 1957, a UNESCO reconnaissance mission suggested that the Galapagos could become an important asset for the Ecuadorian economy by attracting tourism. The 1966 Snow and Grimwood Report recommended ways in which tourism could be managed by large companies (Cairns 2011). The use of cruise boats that took wealthy westerners from island to island was to play a key role in the process (Grenier 2007; Cairns 2011). Currently there are more than 220,000 tourists visiting the Islands every year, and around half of them spend time on the cruise boats. During the last decades, the number and the size of the large cruise boats have increased, and boats have more luxurious and better accommodations. The discourse produced by many of these operators and agencies emphasizes the Galapagos as a pristine land where people are absent and untouched nature can be observed and studied (Grenier 2007; Quiroga 2009).

Three important transformations have occurred over the last years in the tourism industry on the Galapagos: one is that proportionally the number of land-based tourists has increased; secondly that the rotation of the tourist has increased, meaning that tourists stay a shorter time in the archipelago; and thirdly that the number of Ecuadorian tourists has also increased significantly (Epler 2007; Pizzitutti et al. 2016). The threats of an uncontrolled growth in the number of visitors to the Islands have become a major concern. In 2007, the number of tourists exceeded 170,000 according to figures given by the National Park. That same year UNESCO placed the Galapagos on the list of endangered World Heritage Sites, although the Galapagos was removed from the UNESCO endangered list 3 years later, the number of tourists and the threat of new introduced species keeps growing.

There is an important concentration of wealth in the tourism industry in the hands of outsiders. Taylor et al. (2006) have indicated that in 2005, foreigners and mainland residents owned most of the top level luxury boats (almost 82% of them), while Galapagos residents owned only 18%. On the other hand, Galapagos residents owned most of the economy class boats (73%). With some notable exceptions, the companies that own and operate the most luxurious boats are based mostly in Quito and Guayaquil or outside the Galapagos (Taylor et al. 2006; Epler 2007). Studies by the Charles Darwin Research Station show that of the US$419 million spent by Galapagos tourists in 2007, only US$62.9 million entered the local Galapagos economy (Epler 2007; Watkins and Cruz 2007).

Starting in the 1980s, tourists have been staying increasingly in the towns and using the services provided by the local population. This type of land-based tourism has become a major part of the economy in all of the islands (Epler 2007) and has been growing in importance—almost the same number of visitors go to hotels and residencies on land as the numbers that go to the large cruises (Pizzitutti et al. 2016). Land-based operations are now growing faster than cruise boat operations (Quiroga 2013; Pizzitutti et al. 2016). Land-based tourists are a diverse sector that includes most Ecuadorian tourists, some international tourists, backpackers, and international and national students and volunteers. To a large extent, technological improvements and subsidized fossil fuels have allowed the expansion of this land-based tourism. Twenty-four-hour electricity was the result of the introduction of larger diesel electrical generators in the 1980s. Tourism has benefited from an increase in the availability of air conditioners, refrigerators, better digital communication, and other amenities that electricity makes possible, and that makes the towns more attractive to the tourists. Many of the hotels and residencies in the towns are owned by Galapagos residents. The tourists who stay in towns travel from island to island on speedboats and organize daily visits to places close to the ports, often on gasoline-powered outboard engine boats owned and/or operated by fishers (Quiroga 2013; Pizzitutti et al. 2016). Local residents and politicians want more tourists to come to towns, and they claim that what they call the traditional tourism system based on the cruise boats has to change.

Perhaps one of the most dynamic, adaptable, and fastest-growing sectors in the Galapagos is these small- and medium-scale businesses involved in land-based tourism. This sector is not only increasing in economic power but also in its political presence. The land-based tourism involves an increasing number of people including the tour operators, people working for the local hotels, small- and medium-sized operators of overnight boat tours, people working for the speedboat companies who travel between ports or do the island-hopping tours, local travel agencies, guides, and other local people that are increasingly involved, directly or indirectly in land-based tourism. This is a sector that demands electricity for its basic functioning. The rapid growth of this sector, plus the fact that the length of stay of tourists has decreased, explains to a large extent the recent increase in the number of tourists coming to the Galapagos every year in spite of the fact that the total number of cruise boats and berths on liveaboard cruises has not increased significantly. These changes have also meant that there are now more flights and more transportation between islands as port-to-port and island-hopping operations have expanded.

Migration, Population Increase and a Higher Standard of Living

Growth in fisheries and tourism has meant an increase in the number of migrants arriving to the Galapagos. From 1974 to 1982, the annual population growth rate was 4.9%; it increased to 5.9% from 1982 to 1990, and it was 6.7% from 1990 to 1998 before the rate decreased from 2000 to 2010 to 3.3% (Kerr et al. 2004; Villacis and Carrillo 2013; Granda and Salazar 2013; Pizzitutti et al. 2016). The expansion in the 1980s and 1990s of fishing and tourism was one of the main causes of population growth in the Islands as migrants arrived as a response to the demand for labor in fisheries, the tourism sector, or as construction workers. To curb the amount of people migrating to the islands, a law was passed in 1998 regulating who can stay permanently in the Islands. Despite the passing of a law, and a decrease between 2001 and 2010 of the birth rate from 22.7 births per 1000 inhabitants to 14.1 per 1000 inhabitants (Villacis and Carrillo 2013), the number of people living in the Galapagos will reach at least 31,000 people by 2020 (Pizzitutti et al. 2016). The main reason for the population increase has been that jobs in the Galapagos are easier to obtain and are better paid (Villacis and Carrillo 2013). Although the cost of living in the Galapagos is about 80% higher than the rest of the country, it is compensated by the fact that on average, most people receive between 80 and 100% more payment than in the mainland.

Due to higher income, consumption of electricity is higher in the Galapagos than in the rest of the country. Galapagos has one of the highest standards of living and highest rate of consumption of any province of Ecuador (Taylor et al. 2006; Villacis and Carrillo 2013; Sampedro 2017). Rising numbers of ships coming from the mainland with cargo and increased access to goods and services are part of the new Galapagos. Both the number of flights and of cargo ships have surged in the last decades. Airplanes and ships bring home appliances and vehicles that increase the demand for electricity and gasoline on the part of the local residents. According to INEC, in 2009, 93.9% of the families had color TVs, and 30.3% had access to cable TV (INEC 2006). According to the national census office, cell phone coverage is much higher in the Islands than in the rest of the country. 76% of the people had a cell phone in the whole of Ecuador, but 92% had a cell phone in the Galapagos. Similarly, 26% of the households had a computer on the mainland, 47% had one in Galapagos (Villacis and Carrillo 2013). On a survey conducted by INEC, 83.1% of the local people claimed that they would like the number of flights to and from the mainland to increase, 67.3% would like to see more air transportation between islands, and 64.6% would like an increase in the number of boats bringing goods from the mainland (INEC 2006). See Fig. 1.1 for projected increase in round trip airline flights.

Around 87% of the cargo that goes to Galapagos arrives on ships and the rest by airplane. It takes around 3 weeks for a ship coming from Guayaquil to deliver its goods, and their capacity ranges from 300 to 1100 tons (Zapata, 2005; Zapata and Martinetti 2010). Around 7% of the total maritime cargo consisted of food products.

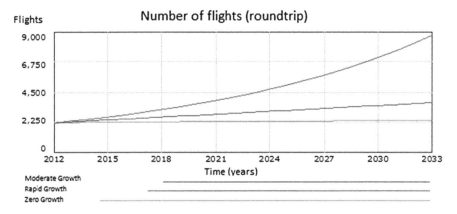

Fig. 1.1 Projected increase in Galapagos airline flights

During the early 2000s, there were five ships serving the Galapagos, but some of them sunk between 2015 and 2016 which produced a deficit of food and cargo among the local residents. Cargo ships arrive to the three main inhabited islands of the archipelago: San Cristobal, Santa Cruz, and Isabela. One of the main problems with the transport of goods to the Galapagos is the inadequate infrastructure of the ports to handle products like fruits and vegetables. In an effort to improve the situation, food started to be transported in containers, but now it is clear that this is not necessarily the best solution either. In 2010 there were 56.142 tons imported for a population of some 28,659 inhabitants (WILDAID 2012).

The increase in the standard of living and consumption level is also reflected in the increase in the number of motorized vehicles in the islands. The number of land vehicles increased between 1998 and 2002 from 28 in 1980 to 1276 in 2006 (Villa 2007). In 2013 the Galapagos government (Consejo de Gobierno del Régimen Especial de Galápagos) established that the ideal number of vehicles for the island was 1932. In 2005 a moratorium was established that limited the number of vehicles that could be brought to the Islands, and a commission was created to control the process. In 2009 the census identified a total of 1962 territorial vehicles in the Galapagos in the 5 populated islands; the largest number of vehicles is on Santa Cruz (1074), followed by San Cristobal (699), Isabela (154), Baltra (24), and Floreana (11) (Oviedo et al. 2010). Most of the vehicles were categorized for personal use (1144), followed by commercial use (181). Motorcycles and scooters were the most common types of vehicles (935), followed by pickup trucks (644) (Oviedo et al. 2010). Figure 1.2 shows three scenarios of future vehicle use in the Islands. Many local people have their own car, and there is a constant use of "taxis," usually white pickup trucks, used in the towns of the Galapagos and represent 20% of the land motor vehicles; 60% of the population uses them more than once a week (Cléder and Grenier 2010). The use of taxis is a very inefficient way of transporting people. Most taxis are pickup trucks that often carry one or two persons without any cargo. Engines are often left running while the drivers or passengers do errands. Many institutions have also pickup trucks that are used to carry people a few blocks

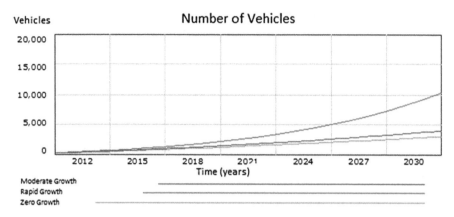

Fig. 1.2 Projected increase in vehicles on the Galapagos Islands. Mena et al. (2013)

in places like Puerto Baquerizo or Puerto Villamil. Bikes on the other hand are not used extensively.

The number of boats taking passengers from one island to the other has also increased steadily. According to the Ecuadorian Navy, there were 42 launches that offered regular transportation between islands in the Galapagos (Ouvrard and Grenier 2010). The size and power of outboard engines have been increasing. Whereas previous to 2000, most engines were only 50 or 75 HP, today most of the boats have 2 or 3 large 150 to 220 HP engines. Boats today have on average 450 HP, and 55% of the boats have reported that they increased the size of their engines in the last few years. They consume an average of 60 gallons per trip between Santa Cruz and Isabella or San Cristobal (Ouvrard and Grenier 2010). The use of more powerful engines has also decreased the average time that a trip between islands takes, which is now 2 h when only a few years ago, it took some 5 h (Ouvrard and Grenier 2010). The most common routes connect two of the three main port towns, like Santa Cruz and Isabela or Santa Cruz and San Cristobal. There are also chartered boats contracted by residents, tourism agencies, or institutions. Many are dedicated to tourism activities and take people to sites close to towns, to SCUBA trips, and to sport fishing outings. As can be seen in Fig. 1.3 in a scenario of rapid growth, the number of inter-island vessels could grow three times with an increase of tourist coming to the Galapagos.

Energy Demand and Supply in the Galapagos

The Galapagos has gone through several systemic transformations (González et al. 2008) during the last 400 years of human presence in the Islands. These transformations have been driven by the increased availability of energy and the constant introduction of technological innovations. Since the 1960s new means of transportation and other technological innovations have resulted in and are the result of population increase and better standard of living in the Islands. These innovations include

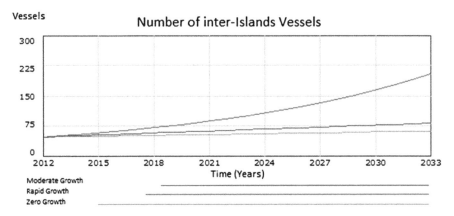

Fig. 1.3 Projected increase in inter-island marine travel. Mena et al. (2013)

airplanes, boats, cars, trucks, outboard engines, refrigerators, and electrical power plants which create an increasing demand for energy. Fossil fuels have made possible the conditions that have accelerated connectivity with the mainland and the world through the expansion of tourism, the increase in fisheries, and the increased level of consumption by the local population. Fossil fuels have provided the bases for greater accessibility of outsiders to the islands and of islanders to outside goods but have also generated conditions that made possible the increase in introduction of invasive species and the risk of major oil spills.

As the need for imported fossil fuels increases, so do the chances of an accident during the transportation of these fuels. One of the most catastrophic events in the history the Galapagos was the oil spill of the fuel tanker "Jessica" the evening of 16 January 2001. The tanker crashed against the rock reef at the entrance to Puerto Baquerizo Moreno, in Wreck Bay, on San Cristóbal Island, Galápagos. The ship was carrying about 600 tons (160,000 gal) of diesel oil and 300 tons (80,000 gal) of intermediate fuel oil (IFO 120). The diesel was destined to be delivered to the fuel dispatch station on Baltra Island, and the IFO was destined for the tourist vessel "Galapagos Explorer" (Sanderson et al. 2001). By 29 January, most of the remaining 180,000 gallons of fuel oil (75,000 gallons of IFO 120 and 105,000 gallons of DO#2) had escaped from the hull and dispersed to waters within the archipelago (Lougheed et al. 2002). Of the 370 large animals reported to be contaminated by oil, most were marine iguanas, pelicans, and sea lions. Other seabirds were also reported, but their numbers were small. Thousands of fish and invertebrates that live in the intertidal and the coastal areas were also affected, but exact numbers were not recorded. The largest numbers of affected animals were found on San Cristóbal and Santa Fe and were reported shortly after the spill (Lougheed et al. 2002). The Jessica oil spill highlighted the need for better system of hydrocarbon transportation and storage. Some improvements such as the use of double hulls and better storage facilities were implemented. It also highlighted the growing need for the introduction of renewable energy. A plan was created a few years later to shift to renewables by 2017 and stop the import of fossil fuels.

Diesel consumption has been growing at rates of 6.00% per year over the last decade. There are different uses for diesel in the Galapagos: in 2001, 60% of the energy went to the tourism sector, 26% to the generation of electricity, 8% to institutions, 4% for fishing, and 2% for transportation in the Islands (Kassels 2003; Langer 2012). In other words most of the diesel that comes to the Galapagos is consumed on the cruise boats, not on land. In the case of gasoline, 41% was used for transportation, 31% used for fishing, 23% for tourism, and 5% for institutions. When taking into account demand for diesel and gasoline, the tourism sector accounted for 41.5%, followed by transportation with 21.5%, fishing 17.5%, the generation of electricity 13%, and used by institutions 6.5% (Kassels 2003; Langer 2012). The demand of fossil fuels used for fishing is decreasing percentagewise, but the demand for other uses such as tourism and land transportation has increased. This large amount of energy that the cruise boat sector consumes coincides with the view of some local residents that large boats use many of the resources and are responsible for many environmental problems.

If we take into account the energy used in the Galapagos, even those fossil fuels used by cargo boats and airplanes that are fueled in the mainland, transportation accounts for the greatest demand for energy: 25,000,000 L or 900,000 GJ went to air transportation; 16,380,000 L or 590,000 GJ went to the large cruise boats; 5,950,000 L or 220,000 GJ went to car transportation; 1,300,000 L or 47,000 GJ went to speedboats; 400,000 L or 14,000 for ferries; and 75,000 L or 27,000 GJ went to cargo boats bringing goods from the mainland (Calle 2014). The majority of the Islands energy demand (80%) is used to transport goods and people in planes, cruisers, and vehicles (Calle 2014) both from the mainland and between islands. Much of the rest goes to the production of electricity, where the system is still relatively inefficient. For example, 170,000 L used for electricity each year, generates 234,400 GJ, of which 60% is lost and 90,000 GJ reaches consumers. GLP is used mostly for cooking (85%) and for heating water for showers and sinks. Gas in Ecuador is commercialized in cylinders of 15 kg for domestic use and of 45 kg for industrial use and are brought in by cargo boats. Just in Santa Cruz between January and August of 2011, a total of 41,000 cylinders were consumed of which 3,100 were for domestic use and 10,000 for industrial use (Calle 2014).

The increased demand for electricity is to a large extent the result of the tourism boom and the way it has stimulated the economy of the Islands. Technological improvements, digital communication, refrigeration, and other goods have become necessary for the evermore demanding tourists arriving to the Galapagos. Electric demand increases during warm season between January and May when there is more demand for energy from cooling systems such as air conditioners and refrigerators.

Energy Subsidies in the Galapagos

The constant growth of the population and the increased consumption of electricity have been accompanied by an increased demand for production of electricity. Much of the electricity produced is from diesel-based thermoelectric generators, but some is based on renewable sources including wind, solar, and vegetable oil.

Fuel subsidies are a big incentive for people to overuse land and marine vehicles. Most of the fuels and the electricity used in the Galapagos are subsidized. In the Galapagos, as in the rest of the country, fossil fuels are subsidized because of social and political reasons, and a gallon of diesel that goes to land diesel vehicles is about four times less expensive than the international prices, and cooking gas is seven times cheaper. Diesel for the tourism sector, which was also subsidized for many decades, is no longer subsidized and tourism companies must pay the international price for the fuel. Besides the national subsidies to fossil fuels, in the Galapagos the transport of fuels is also subsidized as gasoline and diesel cost the same as in the mainland. The price of fuels for artisanal fishers and transport units is $ 1.48 USD for gasoline and 1.02 USD per gallon for diesel. Gasoline for land vehicles and some of the boats cost around $1.60, and the diesel for buses and trucks is $1.40 per gallon. The cost of fuel transportation to the Galapagos in 200 was subsidized 517,000 USD (see Table 1.1).

These fuel subsidies helped support a booming fishing fleet in the 1990s and early 2000s but also contributed to the overexploitation of fish, sea cucumbers, and lobsters. Most of the subsidies were directed to diesel fuel which accounts for five times more subsidies than gasoline. Diesel for electricity generation is also subsidized 0.91 per gallon (Jácome 2008). The amount of money going to subsidize hydrocarbons in the islands has actually increased. In only 5 years, between 1995 and 2000, there were more than 3 million dollars in fuel subsidies in the Galapagos (Kerr et al. 2004). Between 2001 and 2008 subsidies went from accounting 21% of the total price of the fuels to be 58%. Electricity is also heavily subsidized in the Galapagos, it has been calculated that the total amount of subsidies coming to the Islands only for electricity is 8.6 million dollars a year (Jácome 2008; Kerr et al. 2004) calculated that of all the provinces of Ecuador, Galapagos has the highest subsidy.

Ecuador stopped subsidizing fuel in 2012 for national flights, but most of the other subsidies are still in place. Galapagos residents pay around 50% the cost of the ticket to go to the mainland; Ecuadorian nationals pay about 75% of the cost of the ticket, whereas foreigners pay the full cost. Galapagos residents have been receiving traveling subsidies since 1997 (Kerr et al. 2004). Using this subsidy many people fly to the mainland for medical reasons, to go shopping, visit relatives, or for holidays. In a census conducted in 2006, 57.4% of the residents said that they had traveled to the mainland during the last year (INEC 2006).

Table 1.1 Galapagos fuel subsidies in 1996, 1998, and 2000

Year	Gasoline	Diesel	Gas	Estimated transport cost		Total gl	Total tons	Total subsidy
				per gall on	per ton			
	Thousand gl	Thousand gl	Ton	Thousands $	Thousands $	Thousands $	Thousands $	Thousands $
1996	703	2498	528	0.17	181	435	95	530
1998	738	3490	458	0.08	141	279	64	344
2000	1173	4582	379	0.11	82	486	31	517

Source: Petroecuador, taken from Kerr et al. (2004)

As it is true in many places around the world, political pressure makes it very difficult for these subsidies to be removed once they have been implemented. However some of the subsidies, especially those that benefited the large tourism companies, have been removed. In November 2010, Presidential Decree Number 175 eliminated diesel subsidies for large tourism boats and established a formula to calculate the price for the fuels used by the tourism sector. Business people associated with the cruise boats claimed this change represents at a 21% price increase.

Searching for Solutions

In 2007, the government of Ecuador through the Ministry of Renewable Energy outlined a plan called "The Galapagos Islands Zero Fossil Fuel Initiative" to eliminate the consumption of fossil fuels in the Galapagos by the year 2017. It was a plan that aimed at replacing fossil fuels with solar energy, wind energy, and biofuels, and gradually converting diesel engines used in transportation into biofuel engines and using only hybrid cars in the Islands (Carvajal 2012). Although the plan failed as it was initially programed, many important initiatives have been implemented in the last 10 years. Most renewable project initiatives have some financing from the Ecuadorian government and other international cooperation agencies, large utility companies, and government development agencies from countries including Korea, Germany, Japan, and Spain.

The first place where a renewable energy system was successfully implemented was San Cristobal Island. Wind power generation started functioning in San Cristobal in 2007 and provides around 30% of the energy needs of the Island. It consists of three wind turbines each with a capacity to produce 800 KW, producing a total of 3.2 GWh/year. The project was funded by the government of Ecuador, an international consortium of electric companies, and the Galapagos public electric company ELECGALAPAGOS. Between October 2007 and June 2016, diesel generated 71,886,243 kWh (72.4%) of electricity and wind 27,427, 243 (27.4%) (www.eolicsa.com.). Between 2007 and 2015, three 157-foot wind turbines replaced 2.3 million gallons of diesel fuel equivalent to 21,000 tons of carbon dioxide emissions (http://www.elecgalapagos.com.ec, Procopiou 2016). There have also been some technical problems with wind generation resulting in large power fluctuations in San Cristobal affecting some 7000 people. The turbines were going to be originally placed in Cerro Joaquin, but because of fears that they were on the flight route of the Galapagos petrel (*Pterodroma phaeopygia*)—one of the most endangered endemic birds of the islands—they were moved to another hill, Cerro Tropezon. At the moment, the main environmental concern is the risk to bats from the propellers.

Another wind project has been built in Santa Cruz designed to provide 20% of the energy needs of the Island. It consists of a wind park of three turbines on Baltra, an Island next to Santa Cruz, from where the electricity is transmitted to the main port town of Puerto Ayora, where some 14,000 people live. The 50 km transmission line consists of submarine and buried cable. The project can generate 2.25 MW

6000 MWh/year based on three wind turbines of 750 kW each (http://www.elecgalapagos.com.ec). Additional photovoltaic projects have been built on Santa Cruz and Baltra. In Puerto Ayora a 1.5 MWp solar farm has been built on 2.9 hectares of land. It consists of 6000 solar panels and was financed by the Ministry of Electricity of Ecuador, the Electric Energy Company of Galapagos, and the Korean cooperation agency KOICA. The photovoltaic project in Baltra produces 0.2 KWp, and its aim is to stabilize the energy generated by the wind turbines on the Island. It was financed by Japan Cooperation Agency (JICA) and the Ministry of Electricity and Renewable Energy (MEER). In Floreana the original 2004 solar energy project stopped working in 2009, but has been renewed and supplemented with a bio-oil project based on a Jatropha plant. There is now 21 kWp of PV energy installed as of 2014 based on funding from the AECI (Spanish Cooperation Agency) and the GIZ (German Cooperation Agency).

Jatropha (*Jatropha curcas*) is a native plant grown on the coast of Ecuador where it has traditionally been used for many purposes but especially as a fence. Thermic plants using Jatropha oil have been built on Isabela and Floreana. Large amounts of oil are needed to produce fuel for the two islands: 30,000 gal/year in Floreana; 180,000 gal/year in Isabela (Jácome 2008). The project consists of having farmers in the province of Manabí grow the plant and then process it into oil for delivery to the Galapagos. This project can create jobs in rural areas of the mainland without taking away agricultural land as it grows in marginal soils.

In Isabela, an island that has a population of more than 2000 people, there are two solar photovoltaic generators that produce electricity on the Island. These generators have a total capacity of 800 KWpv and improving the existing hydrothermal generators will reduce the amount of diesel imported to the Island by around 60% annually (status 2007), which means a reduction of 400 L of diesel per day (more than 146.000 L of diesel per year). (ELECGALAPAGOS S.A. N.D. report). Working with the German Development Bank and the Ministry of Electricity and Sustainable Energy of Ecuador, a solution has been proposed for Isabela that includes several strategies: a thermic plant fueled by a mix of Jatropha oil and diesel that will produce 1.62 MWp, a solar photovoltaic plant that can generate 0.92 MWp, and an energy storage system based on lithium-ion batteries and storage tanks for the Jatropha oil (http://www.elecgalapagos.com.ec).

At a local scale, different solutions have been proposed, and some have been implemented to lower the dependency on fossil fuels. Municipalities and small business have used solar panels to generate electricity for buildings and street lights. Besides the production of alternative energy, projects to decrease energy consumption are also in place and include banning of cars, light bulb changes, and a new type of induction cooking stove. Another project replaced refrigerators older than 10 years for new ones. 739 refrigerators have been replaced up to June 2014 representing a reduction in energy use of 7.5%. There has also been an effort to replace conventional public lighting with more efficient lamps LED. Bike paths have also been created in most of the inhabited islands by the municipalities.

Although there has been an increase in the number of renewable energy projects, the relation between renewable energy and fossil fuels is not getting better because of the increased demand for energy. Currently the amount of electricity that is generated from alternatives is still relatively small. However according to some projections (Pizzitutti et al. 2016), if new renewable energy projects are not built, and demand for electricity continues to increase, the ratio of renewables to traditional fuels will decrease over time.

Conclusion

The unique and fragile ecosystems of the Galapagos depend on a high degree of isolation for the islands. The relatively recent presence of humans in the last few centuries has generated a series of impacts and transformations on the environment and on the iconic species. During the last century, technological changes and the availability of cheap fossil have generated a positive feedback system which results in an increase in the human presence, an expansion of the economy based mostly in tourism, and an increase in the consumption of goods on the islands. The international fame of the Galapagos as a pristine natural laboratory has resulted in the increase of the number of cargo boats and airplanes coming and going to the islands. The increased amount of traffic has resulted in an increase of alien species being introduced to the islands. Currently there are more introduced species of plants than there are native or endemic species, and the number of new introduced insects increases every year and is a threat to Galapagos ecosystems. The increase in marine traffic has also created a greater risk for oil spills that could affect the fragile marine and intertidal communities of the Galapagos.

Island-hopping tours in speedboats and other land-based tourist activities have started to demand more fossil fuels as outboard engines, air conditioners, refrigerators, and digital communications have been introduced at a fast pace to satisfy the demand of the local population and the tourists. Fossil fuels sold in the Galapagos are subsidized; which has contributed to the increases in the standard of living of the local population. Subsidized fossil fuels are a challenge to the conditions necessary to support the evolutionary processes needed for the maintenance of the Galapagos.

In order to decrease dependency on fossil fuels, Galapagos has experimented with a range of energy types and is trying to develop new energy mixes. The vulnerability of the animals and the ecosystems of the Galapagos has been a critical factor in creating a sense of urgency about generating the proper mix. Some of the solutions so far have been directed to the uses of alternative energy by the local residents, but there are still many improvements that can be made to lower energy consumption. Small changes in patterns of consumption and energy use such as increased use of bicycles and less use of appliances such as air conditioners together with improvements in design and construction can help to improve energy management in the Islands. There have been few if any programs aimed at improving the efficiency of energy use in transportation. Realistic solutions to reduce the use of fossil fuels in sea and air transportation seem to be distant. A combination of biofuels, solar energy, and the use of sail boats could help reduce the dependence on nonrenewable sources in transportation. The positive feedback system that generates higher levels of growth and development, threatens local and iconic organisms and the unique and fragile ecosystems of the Galapagos. Fossil fuels are a challenge to the conditions that support the evolutionary process necessary for the maintenance of Galapagos as a natural laboratory. New technologies need to be developed to minimize hydrocarbon fuels currently used in transportation of goods and tourists. Currently, the system is based on the demand of tourists to see many places in a few days. As a major use of fossil fuels is traveling to and from the mainland and with no alternatives to the use of airplanes, the rotation of tourists and the time spent by tourists on the islands should be re-examined.

References

Ben-Yami M (2001) Managing artisanal fisheries of Galápagos. A consultancy report—07-01-2001–03-02-2001. Charles Darwin Research Station, Santa Cruz, Galápagos

Buddenhagen CE (2006) The successful eradication of two blackberry species Rubus megalococcus and R. adenotrichos (Rosaceae) from Santa Cruz Island, Galapagos, Ecuador. Pac Conserv Biol 12(4):272–278

Cairns R (2011) A critical analysis of the discourses of conservation and the science of the Galapagos Islands. PhD dissertation, University of Leeds, UK

Calle P (2014) MA thesis, Delft University of Technology Faculty Architecture Department Building Technology Studio Sustainability Studio

Carvajal P (2012) Galapagos islands zero fossil fuel initiative. Ministry of Electricity and Renewable Energy-Ecuador. Retrieved from http://www.irena.org/DocumentDownloads/events/MaltaSeptember2012/Pablo_Carvajal.pdf

Causton CE, Peck SB, Sinclair BJ, Roque-Albelo L, Hodgson CJ, Landry B (2006) Alien insects: threats and implications for conservation of Galapagos Islands. Ann Entomol Soc Am 99(1):121–143

Cléder E, Grenier C (2010) Taxis in Santa Cruz: uncontrolled mobilization CDF, GNP, and Governing Council of Galapagos, 2010. Galapagos Report 2009–2010. Puerto Ayora, Galapagos

Engie K (2014) Adaptation and shifting livelihoods of small-scale fishers in the galápagos marine reserve. Ecuador PhD. Dissertation University of North Carolina Chapell Hill

Epler B (2007) Tourism, the economy, population growth, and conservation in Galápagos. Santa Cruz, Galápagos, Charles Darwin Foundation, p 68

Gardener MR, Trueman M, Buddenhagen C, Heleno R, Jäger H, Atkinson R, Tye A (2013) A pragmatic approach to the management of plant invasions in Galapagos. In: Plant invasions in protected areas. Springer, Dordrecht, pp 349–374

González JA, Montes C, Rodríguez J, Tapia W (2008) Rethinking the Galápagos Islands as a complex social-ecological system: implications for conservation and management. Ecol Soc 13(2):13

Granda L, Salazar M y GC (2013) Población y migración en Galápagos. En: Informe Galápagos 2011–2012. DPNG, GCREG, FCD y GC, Puerto Ayora, , Galápagos, Ecuador, pp 44–51

Grenier C (2007) Conservación contra natura. Las islas Galápagos (Vol. 233). Editorial Abya Yala

Guézou A, Chamorro S, Pozo P, Guerrero AM, Atkinson R, Buddenhagen C, Jaramillo Díaz P, Gardener M (2016) CDF Checklist of Galapagos Introduced Plants—FCD Lista de especies de Plantas introducidas Galápagos. In: Bungartz F, Herrera H, Jaramillo P, Tirado N, Jiménez-Uzcátegui G, Ruiz D, Guézou A, Ziemmeck F (eds) Charles Darwin Foundation Galapagos Species Checklist - Lista de Especies de Galápagos de la Fundación Charles Darwin. Charles Darwin Foundation/Fundación Charles Darwin, Puerto Ayora, Galapagos. http://darwinfoundation.org/datazone/checklists/introduced-species/introduced-plants/. Last updated: 29 Sep 2016

Hearn A (2008) The rocky path to sustainable fisheries management and conservation in the Galapagos Marine Reserve. Ocean Coast Manage 51(8–9):567–574. https://doi.org/10.1016/j.ocecoaman.2008.06.009

INEC (2006) Censo de Población y Vivienda. Ecuador

Jácome C (2008) Subsidios en el sector energético insular. Informe Galápagos 2006–2007. FCD PNGINGALA, Puerto Ayora, Santa Cruz, pp 67–72

Jaramillo P, Guezou A (2012) CDF checklist of Galapagos vascular plants. Charles Darwin Foundation. http://www.darwinfoundation.org/datazone/collections/. Accessed 9 Dec 2012

Jiménez-Uzcátegui G, Carrión V, Zabala J, Buitrón P, Milstead B (2008) Status of introduced vertebrates in Galapagos. Galapagos Report 2007–2008. Charles Darwin Foundation, PuertoAyora, pp 97–102

Kassels K (2003) Energy evolution: renewable energy in the Galapagos Islands. Refocus 4:36–38

Kerr S, Cardenas S, Hendy J (2004) Migration and the environment in the Galapagos: an analysis of economic and policy incentives driving migration, potential impacts from migration control, and potential policies to reduce migration pressure. Motu Working Paper 03–17, Motu Economic and Public Policy Research, Wellington

Langer K (2012) Potential of renewable energies in the Galapagos archipelago-acceptance of technological systems in a protected environment. M.A. dissertation, Universidad Autónoma de San Luis Potosí Facultades De Ciencias Químicas, Ingeniería y Medicina Programas Multidisciplinarios de Posgrado En Ciencias Ambientales and Cologne University of Applied Sciences Institute For Technology and Resources Management in the Tropics and Subtropics

Latorre O (2001) La Maldición de la Tortuga: Historias Trágicas de las Islas Galapagos (Cuarta Edi). Artes Grá fi cas Senal, Impresenal Cia, Quito, Ecuador, p 226

Lougheed LW, Edgar GJ, Snell HL (2002) Biological impacts of the Jessica oil spill on the Galapagos environment: final report. Charles Darwin Foundation for the Galapagos Islands

Mena CF, Walsh S, Pizzitutti F, Reck G, Rindfuss R, Orellana D, Granda VT, Valle C, Quiroga D, García JC, Vasconez IL, Guevara A, Sanchez ME, Frizelle B, Tippett R (2013) Determinación de las relaciones sociales, ambientales y económicas que permitan desarrollar, en base a procesos de modelación, potenciales escenarios de sostenibilidad del sistema socio-ecológico de las islas galápagos con énfasis en la dinámica del flujo de visitantes al archipiélago. CDC-MAE-032-2013

Ouvrard E, Grenier C (2010) Transporting passengers by launches in Galapagos CDF, GNP, and Governing Council of Galapagos, 2010. Galapagos Report 2009–2010. Puerto Ayora, Galapagos, Ecuador

Oviedo M, Agama J, Buitrón E, Zavala F (2010) The first complete motorized vehicle census in Galapagos CDF, GNP, and Governing Council of Galapagos, 2010. Galapagos Report 2009–2010, Puerto Ayora, Galapagos, Ecuador

Peck SB (1996) Origin and development of an insect fauna on a remote archipelago: the Galapagos Islands, Ecuador. In: Keast A, Miller SE (eds) The origin and evolution of Paci fi c Island biotas, New Guinea to Eastern Polynesia: patterns and processes. Academic Publishing, Amsterdam, pp 91–122

Pizzitutti F, Walsh SJ, Rindfuss RR, Gunter R, Quiroga D, Tippett R, Mena CF (2016) Scenario planning for tourism management: a participatory and system dynamics model applied to the Galapagos Islands of Ecuador. J Sust Tour, 1–21. https://doi.org/10.1080/09669582.2016.1257011

Procopiou C (2016) The wind turbines saving the Galápagos Islands. Newsweek

Quiroga D (2009) Crafting nature: the Galapagos and the making and unmaking of a "natural laboratory". J Politic Ecol 16(1):123–140

Quiroga D (2013) Changing views of the Galapagos. In: Walsh SJ, Mena CF (eds) Science and conservation in the Galapagos Islands. Springer, New York, NY, pp 23–48

Sampedro C (2017) System dynamics in food security: agriculture, livestock, and imports in the Galapagos Islands. MA Thesis Universidad San Francisco de Quito

Sanderson W, Tiercelini C, Villanueva J (2001) Accident of the Oil Tanker Jessica. Off the Galapagos Islands (Ecuador) Final Report to the European Commission DG. Environment ENV.C.3. OF THE OIL TANKER

Schuhbauer A, Koch V (2013) Assessment of recreational fishery in the Galápagos Marine Reserve: failures and opportunities. Fish Res 144:103–110

Simberloff D (2013) Invasive species: what everyone needs to know. Oxford University Press, New York

Taylor JE, Hardner J, Stewart M (2006) Ecotourism and economic growth in the Galapagos: an island economy-wide analysis. Environ Dev Econ 14:139–162

Tye A (2001) Invasive plant problems and requirements for weed risk assessment in the Galapagos islands. In: Groves RH, Panetta FD, Virtue JD (eds) Weed risk assessment. CSIRO, Melbourne, pp 153–175

Tye A (2008) The status of the endemic flora of Galapagos: the number of threatened species is increasing. Galapagos report 2007–2008. Charles Darwin Foundation, Puerto Ayora, pp 97–102

Valle CA (2013) Science and conservation in the Galapagos Islands. In: Walsh SJ, Mena CF (eds) Science and conservation in the Galapagos Islands. Springer, New York, NY, pp 1–22

Villa A (2007) El crecimiento del parque automotor en Galápagos. In: FCD, PNG & INGALA. Informe Galápagos 2006–2007. Puerto Ayora, Galápagos, Ecuador

Villacis B, Carrillo D (2013) The socioeconomic paradox of Galapagos (2013). In: Walsh SJ, Mena CF (eds) Science and conservation in the Galapagos Islands. Springer, New York, NY, pp 23–48

Watkins G, Cruz F (2007) Galápagos at risk: a socioeconomic analysis of the situation in the archipelago. Charles Darwin Foundation, Puerto Ayora, Galápagos

Wildaid (2012) La cadena de cuarentena: Estableciendo un sistema eficaz de bioseguridad para evitar la introducción de especies invasoras a las Islas Galápagos. San Francisco, CA

Zapata F (2005) Diagnóstico del sistema óptimo de transportación marítima de carga hacia Galápagos. Programa de las Naciones Unidas para el Desarrollo (UNDP), Puerto Ayora, Santa Cruz, Galápagos, pp 15–57

Zapata F, Martinetti M (2010) Optimizing marine transport of food products to Galapagos: advances in the implementation plan. Galapagos Report 2009–2010. Puerto Ayora, Galapagos

Chapter 2
Driving Forces and Barriers for a Sustainable Energy Mix in Fragile Environments: North–South Perspectives

Michael S. Quinn

Introduction

The 133,000 km^2 Galapagos Marine Reserve, established in 1998, is one of world's most biological diverse protected areas. Equally significant is the fact that approximately 97% of the land area of the Galapagos Islands is protected as a national park and the entire region is designated a UNESCO World Heritage Site. More than 3000 species, 20% of which are endemic, make this one of the world's most precious and beloved biodiversity hotspots. Due to the isolation and protection of the Galapagos, approximately 97% of the original flora and fauna composition remains intact.

The unique environment of the archipelago draws an exponentially increasing number of nature-based tourists. Meeting the demands of these visitors taxes the capacity of the landscape to meet the energy requirements and assimilate the impacts. This is a remote, fragile and highly cherished environment where decisions about human use and management are paramount. Although only 3% of the land area is inhabited, there are a myriad of effects arising from human presence. As with other fragile environments, we are 'loving the Galapagos death'.

The conveyance and use of diesel for thermal energy and transportation represent one of the greatest threats to this fragile island ecosystem. The Galapagos Islands imported 47,585,127 L of diesel and gasoline in 2010 (Westerman 2012). Oil is shipped at frequent intervals by relatively small tankers from mainland Ecuador to limited storage facilities on the islands. Until 2007, nearly all the electricity used in the Galapagos was derived from diesel generators. Diesel is also the primary fuel for marine vessels. In January 2001, risk became reality when the tanker 'Jessica' ran aground 1 km off San Cristobal Island spilling three million litres of bunker C (intermediate fuel oil 120) and diesel oil into Wreck Bay. The potentially cata-

---

M.S. Quinn (✉)
Mount Royal University, Calgary, AB, Canada
e-mail: mquinn@mtroyal.ca

© Springer International Publishing AG 2018
M.-E. Tyler (ed.), *Sustainable Energy Mix in Fragile Environments*,
Social and Ecological Interactions in the Galapagos Islands,
https://doi.org/10.1007/978-3-319-69399-6_2

strophic ecological effects were partially abated by weather and sea conditions when some of the oil evaporated and much of it dispersed over a large area (Edgar et al. 2003). Nonetheless, there were negative consequences recorded for iconic species such as the marine iguana (Lougheed et al. 2002), and the clean up/restoration costs were approximately US $9 million. The event served as a significant driver for considering alternatives to an energy system dominated by fossil fuels. In 2007, both the Ecuadorian government and UNESCO declared the Galapagos as 'threatened' and in need of priority action (González et al. 2008).

Energy, Climate Change and Tourism on Remote Island States

Along with the potential for oil spills, remote island states are highly susceptible to changes in global climate as a result of fossil fuel consumption. For example, Simpson et al. (2010) reported that islands of the Caribbean are:

> particularly vulnerable to the effects of climate change, sea level rise and extreme events, including: relative isolation, small land masses, concentrations of population and infrastructure in coastal areas, a limited economic base and dependency on natural resources, combined with limited financial, technical and institutional capacity for adaptation. (p. 4)

The authors estimate that a conservative estimate of 1 m rise in sea level in the next 50 years could result in the following losses in the Caribbean:

- Nearly 1300 km^2 land area lost (e.g. 5% of the Bahamas, 2% Antigua and Barbuda)
- Over 110,000 people displaced (e.g. 5% of population in the Bahamas, 3% Antigua and Barbuda)
- At least 149 multimillion dollar tourism resorts damaged or lost, with beach assets lost or greatly degraded at many more tourism resorts
- Damage or loss of five power plants
- Over 1% agricultural land lost, with implications for food supply and rural livelihoods (e.g. 5% in Dominica, 6% in the Bahamas, 5% in St. Kitts and Nevis)
- Inundation of known sea turtle nesting beaches (e.g. 35% in the Bahamas and St. Kitts and Nevis, 44% in Belize and Haiti, 50% in Guyana)
- Transportation networks severely disrupted:
    - Loss or damage of 21 (28%) CARICOM airports
    - Lands surrounding 35 ports inundated (out of 44)
    - Loss of 567 km of roads (e.g. 14% of road network in The Bahamas, 12% Guyana, 14% in Dominica)

The above changes are predicted to result in direct rebuilding costs to the Caribbean tourism industry of US $10 to $23 billion by 2050.

Similarly, climate change is expected to exacerbate and intensify the effects experienced during El Niño episodes in the Galapagos leading to increased temperatures, changes in precipitation, seal level rise, ocean acidification, and ocean current shifts, all of which are highly detrimental to regional biodiversity. Moreover, tourism is responsible for at least 75% of the economy in the Galapagos and directly employs approximately 40% of the islands' population:

> Climate change is expected to threaten all of the species that research shows are most important to tourists. Severe declines in these species could lead to either a reduction in tourism or a shift from nature-based tourism to more mass-market, resort-based tourism. Such a shift would further threaten wildlife species, as this style of tourism likely would require additional urban development and natural resources and result in increased habitat loss and pollution. (Conservation International and WWF 2011, p. 12)

Despite repeated warnings and calls to limit the number of tourists, annual visitation continues to grow at a rate of approximately 14% per year; this equates to a doubling of tourists every 5 years. Tourism began to increase in the 1960s when approximately 1000 visitors/year travelled to the islands. In 1990 there were approximately 40,000 visitors to the Galapagos, and today the number exceeds 200,000 (Tourtellot 2015). The ever-increasing number of visitors creates a demand for 'more': "more electrical power, more food, more water, more fuel, and a larger workforce: and thus more immigration, more garbage, more transportation, more introduced species and greater danger to sustainability" (Sevilla 2008, p. 27). "Energy flows, growing economic opportunities, and positive economic feedback associated with the current model of tourism have the potential to accelerate major changes and threaten the sustainability of the archipelago in the near future" (González et al. 2008, p. 13). The transition to a sustainable energy mix for the islands is essential to the future of the entire social-ecological system. The Galapagos Islands and other fragile environments offer unique opportunities to develop and implement strategies for sustainable, resilient energy systems.

Energy systems are comprised of diverse actors (e.g. producers, transporters, regulators, end-users) with different, often competing, objectives all interacting through physical and social network that are fraught with uncertainty. Bale et al. (2015) argue that adopting approaches and models from complex systems, thinking is necessary to address the challenges of providing affordable, secure energy in a manner that mitigates climate change. "Complexity science and its associated modelling methods enable the study of how interactions between different elements of a system give rise to the collective emergent behaviour of that system and how the system interacts and responds to its environment and evolves overtime" (Bale et al. 2015, p. 151). The purpose of this chapter is to review the critical driving forces and barriers to achieving sustainable energy systems and energy security in fragile environments. Precedents from other fragile and remote regions are offered as sources of potential actions and solutions. This information provides a critical foundation for a complexity theory approach to addressing energy management in the Galapagos Island.

Driving Forces

The preceding introduction clearly establishes *international significance* as one of the primary drivers for achieving a sustainable energy mix on the Galapagos Islands. The region is the focus of international attention, and the global community expects the Ecuadorian government to address the need for change. Since at least 2007, the national administration has been responding to both external and domestic imperatives for a transition to a more renewable energy mix. For example, the Ecuadorian government launched a 'Zero Fossil Fuels in the Galapagos' programme in 2007 which hoped to eliminate the use of fossil fuels in the electrical sector by 2015 and the transportation sector at a later date (IRENA 2015).

Global significance and recognition has resulted in *interest from a variety of international organizations* that engage in collaborative sustainable energy programmes. For example, the Ecuadorian government launched the 'Renewable Energy for Electricity Generation-Renewable Electrification of the Galapagos Islands' programme supported by the Global Environment Facility (GEF) and the United Nations Development Programme (UNDP). Likewise, the presence of groups such as the E-8, a non-profit international organization composed of nine leading electricity companies from the G8 countries, is a potent driver of change. The E-8 helped to develop and implement public-private partnerships that led to wind (2.4 MW) and solar energy (2 × 6 kW) installations on San Cristobal Island. The US $10 million wind project is expected to reduce terrestrial diesel consumption by approximately 50% (E-8 2008), although the system has not met expectations to date (Yu et al. 2015). The project also includes a strong commitment to technical capacity building and public education on sustainable energy. "Through these programmes, the e8 has highlighted the importance of human capacity building and public education for the effective local acceptance, development and spread of renewable and clean energy technologies" (E-8 2008, p. 50).

Another related driver for change is the *expectations of visitors* traveling to the islands to experience the perceived pristine conditions. Tourists willing to travel 1000 km off the western coast of Ecuador are doing so in order to experience some of the highest quality nature-based tourism on the planet. The future of this industry will be based on the ability to insure that the unique flora and fauna of the islands is protected from the effects of human presence, including the consequences of fossil fuel combustion.

Global climate change initiatives are another major driver for addressing energy issues. The Republic of Ecuador submitted a climate change action plan as part of the UN Framework Convention on Climate Change (UNFCCC) negotiated in Paris in late 2015. Ecuador's commitment is based on the principle of *Buen Vivir* (good living – from the Ecuadorian Constitution) – "to live with dignity and have basic necessities met in harmony with oneself, with the rest of the community, with different cultures and with nature" (Government of Ecuador 2015, p. 2). Meeting these commitments is, and will increasingly be, an important driver in the reduction of hydrocarbon combustion. As indicated above, the islands are potentially imperilled by the altered environmental conditions expected to accompany climate change.

Political will and administrative commitment are critical drivers in the establishment of a sustainable energy mix. The actions required to achieve sustainability extend well beyond relatively short-term political electoral cycles. The development of policy, law, regulations and guidelines need to be conducted with intent to grow the commitment over time. This needs to occur at the national level as well as the local level of islands.

Local residents and business owners can also be an impetus for change. People living and relocating to the Galapagos are likely to have a strong desire to maintain the conditions that support their livelihood. Organizations such as the International Galapagos Tour Operators Association (IGTOA) strongly support environmental management for sustainability. "Residents and tourists in the Galápagos are inextricably linked: For locals, keeping the Galápagos pristine is not simply a matter of protecting biodiversity but necessary for the economy, which thrives as a result of the tourist industry" (IGTOA 2016). Developments like Pikaia Lodge on Santa Cruz Island help drive the demand for a lower carbon future on the islands. The lodge is carbon neutral and expects to create a net surplus of energy to put back into the grid (Pikaia Lodge 2016). This is a tourism product aimed at the discerning visitor who expects a high-quality environment and luxury accommodation. Helping to choose the type, timing and intensity of tourism on the islands is a driver that can be strongly influence by local interests.

Barriers

There is no shortage of barriers to a more sustainable energy transition. In 1997 the Global Environment Fund (GEF) project identified the following as some of the most significant barriers to renewable energy alternatives in the Galapagos:

- Limited experience with renewable energy technologies, especially with regard to electricity generation
- Lack of familiarity with the operation and maintenance of renewable energy and hybrid (renewable/conventional) electricity systems
- High initial capital cost of renewable energy technologies
- No experience with power purchase agreements and independent power generation
- Lack of experience with project finance investments and joint venture operations between electric utilities and the private sector
- Difficult access to finance for renewable energy technologies that are new to Ecuador due to high perceived risks

Westerman (2012) reported that the primary drivers of energy consumption for households and businesses in the Galapagos Islands are subsidies ($13,402,294 expended on gasoline and diesel subsidies for the islands in 2005), inefficient electrical appliances and lack of awareness about renewable energy alternatives.

These factors act as barriers to achieving more efficient energy systems and can be addressed by relatively simple policy intervention.

Path dependency refers to the phenomenon of past conditions influencing current decisions. Systems theory asserts that complex systems exhibit 'memory' (hysteresis) whereby current states are sensitive to initial conditions (Shapiro and Summers 2015). Unruh (2000) describes how industrial economies have been locked into fossil fuel-based energy systems "through a combination of systematic forces that perpetuate fossil fuel-based infrastructures in spite of their known environmental externalities and the apparent existence of cost-neutral, or even cost-effective, remedies" (p. 817). Path dependency acts as a barrier to innovation diffusion. The adoption of alternative and renewable energy sources requires overcoming significant inertia of the existing energy system in the Galapagos.

Remoteness is one of the most obvious barriers to sustainable energy transitions in the Galapagos. The islands are located approximately 1000 km west of the Ecuadorian mainland. There are significant challenges and costs associated with the transportation of all products and technologies utilized on the islands. Remoteness is a financial barrier and a disincentive for investors in alternative energy production. The small land area and disaggregated isolation of energy loads further exacerbate the remoteness of the islands from an energy systems perspective due to economies of scale.

The majority of tourism in the Galapagos is facilitated by marine vessels for both transportation and accommodation. These boats are primarily fuelled by diesel that is transported from the mainland and stored on the islands. Despite the growing availability of alternative fuels and technologies (Deniz and Zincir 2016), the *challenges of adapting and retrofitting a fleet* to utilize alternative fuels are a significant barrier to a more sustainable energy future. Likewise, other technical challenges associated with emerging technologies for low carbon energy production, storage and transmission act as barriers throughout the terrestrial environs of the Galapagos.

Political and regulatory frameworks can promote or inhibit the transition to a more sustainable energy system. An historic barrier to the adoption of alternative energy sources is the subsidization of electricity and transportation fuel in the Galapagos. Although this subsidization helps offset the potentially high costs to local consumers, the energy prices do not reflect the real costs of transporting, generating, distributing and addressing the impacts of energy production (Jácome 2007).

Precedents

Barbados

A draft National Sustainable Energy Policy of Barbados (Government of Barbados 2016) was developed to address the heavy reliance on fossil fuels (97.4% of total energy is oil derived). The objectives are to "unlock economically viable investments in sustainable energy that will reduce Barbados' dependency on fossil fuels, and therefore reduce energy costs, improve energy security, and enhance environmental sustainability" (p. 8). The policy is based on a set of core principles that (1) include a commitment to reduce costs and increase environmental sustainability, (2) full-cost accounting (i.e. account for externalities not considered in conventional cost-benefit analyses) and maximize value creation, (3) access international support (including regional integration), (4) include technological neutrality, (5) build on existing strengths of the current utility system and (6) support cost-effective research and development for energy innovation. The policy will be delivered through a set of actions that increase the renewable energy production (29% of all electricity consumption will be from renewable sources by 2029), increase electrical energy efficiency to generate 22% savings over business as usual conditions by 2029, improve non-electric energy efficiency (primarily in transportation) by 29% by 2029, increase the sustainability and efficiency of all fossil fuel activities and increase awareness and skills of the Barbados population for economically viable and environmentally sustainable energy production and use.

An innovative mechanism to help achieve the objectives of the sustainable energy policy in Barbados is the creation of the Energy Smart Fund to provide financial assistance and technical support for the development energy efficiency and renewable energy projects. Capital for the programme was created through a US $10 million loan from the Inter-American Development Bank. The Energy Smart Fund is active in six programmes including (1) technical assistance through grants to businesses for predevelopment studies for energy efficiency and renewable energy projects, (2) financing for energy retrofitting, (3) consumer rebates and financing, (4) free compact fluorescent bulb distribution, (5) rebates for the replacement of old air conditioners and (6) grants for public awareness and education programmes.

The removal of barriers and creation of incentives have demonstrated how an island state like Barbados can improve energy efficiency, security and sustainability. In late 2015 the Caribbean Centre for Renewable Energy and Energy Efficiency (CCREEE, see: www.ccreee.org) was established with Barbados as the regional implementation hub. This will help grow the lessons learned in Barbados across the rest of the Caribbean and beyond.

Canary Islands

The Canary Islands are located 100 km off the southwest coast of Morocco. The Spanish archipelago is comprised of seven main islands including El Hierro (278 km^2, population 10,750) which claims to be one of the world's first energy self-sufficient islands. Similar to the Galapagos, the island harbours a unique and fragile ecological system and is a UNESCO Biosphere Reserve. It previously relied on marine transportation of diesel fuel to power its electrical generators (45 GWh/year from nine units). In 2014 El Hierro transitioned to a combined wind-hydro system (Gorona del Viento) comprised of five 2.3 MW wind turbines (11.5 MW total) backed up by hydroelectric system. A similar system has been proposed for Gran Canaria, the most populous island in the archipelago (Bueno and Carta 2006). During periods when energy production exceeds demand, water is pumped from a reservoir near sea level to a natural volcanic crater approximately 655 m higher. When wind abates, water is released through a set of turbines in a pipe that connects the two reservoirs. The hydro system is capable of producing 11.32 MW at maximum flow. It is estimated that the integrated renewable energy system will result in eliminating annual emissions of 100 tonnes of SO_2, 400 tonnes of NO_x and 18,700 tonnes of CO_2 (Godina et al. 2015) and will meet project energy demands to 2030.

The El Hierro project required a significant financial investment of approximately 64 million Euro; 26% private and 74% public sector sources. The project offers a useful comparison to the Galapagos as the primary driver was sustainable tourism. Not only did the project help to protect the natural environment, the key amenity for tourism, but the facility itself has become an attraction for guided tours and trekking (Sanchez 2014). The project also includes three desalinization plants thus providing critically needed freshwater to inhabitants and limiting the impact on natural aquifers. Finally, not only has El Hierro achieved its sustainable energy targets, it has also achieved energy independence and security for the future. Bueno and Carta (2006, p. 338) conclude that "these systems represent an enormous and as yet barely explored potential" for achieving a more sustainable energy mix in remote and fragile environments.

Northern Canada

Northern Canada (and other peripherally remote areas) is characterized by widely dispersed, remote communities with little or no connection to centralized energy grids. Although they are not physically islands, they are faced with many of the same challenges to achieve greater energy security and sustainability as the Galapagos. The 2006 Canada Census identified 292 remote communities with a total population of nearly 200,000 living 'off the grid' (Natural Resources Canada 2011). The vast majority of these communities (251) rely on diesel generators for the production of electricity with a combined output of approximately

450,000 kW. Most of these communities also rely on heating fuel for their heat and cooking. Although diesel is highly subsidized in these communities, the electricity rates paid by consumers is still three times what is paid by on-grid customers. As with the islands, the drivers for a more sustainable energy mix include environmental factors (greenhouse gas emissions, potential for spills in transportation, impacts of infrastructure and contamination from on-site spills), social factors (noise pollution, health effects, potential for black-outs) and economic factors (high costs of transportation, low incentive for innovation due to economies of scale).

Alternative energy mixes are being considered and deployed in many of these communities. Renewable alternatives include hydro, biomass, geothermal, wind and solar energy. The federal government created 'ecoENERGY' for Aboriginal and Northern Communities Program in 2007 to support the development and implementation of renewable energy projects. The programme has supported nearly 300 community projects ranging from $4300 to $250,000 for a total of $35 million. These funds were generally matched through other sources to generate an even greater impact.

One example from the ecoENERGY programme is a 152 kW photovoltaic solar array at the Deer Lake First Nation Elementary School in northern Ontario. The project was implemented in collaboration with a First Nation renewable energy management company formed by the chiefs of the six communities of Keewaytinook Okimakanak. The company, NCC Development, aims to reduce the reliance on diesel fuel by remote First Nations in northwestern Ontario by 50% through strategic implementation of solar microgrids, energy conservation and load management (NCC 2016). The project enabled five new homes to be connected to the local grid and will meet the school's energy load during daylight hours. The PV system augments a 149 kW run-of-the river hydro system and a diesel power plant.

One of the key lessons learned through the ecoENERGY programme is the importance of relationship building and working with the communities (communications, networking and outreach) to build the necessary capacity for adopting alternative energy production (Indian and Northern Affairs Canada 2010). The move to lower carbon energy mixes is much more than a technical exercise; it is about cultural and societal shifts. The programme also identified the interconnectedness of energy systems with other elements of the social-ecological system. Renewable energy systems need to be integrated with other programmes and administrative structures such as community economic development, strategic planning, community health and education.

Conclusions

The Galapagos Islands are a biodiversity hotspot and world-class tourist destination with ever-increasing visitation. Like other remote, fragile environments, the islands face pressing challenges to meet their energy needs in a secure, economically efficient, socially acceptable and environmentally appropriate manner. The transition to a lower carbon energy future will require exploration of new mixes of energy production in a way that maintains flexibility in the face of change and uncertainty.

Energy systems are highly complex entities comprised of many actors, often with conflicting objectives, operating within dynamic physical, social, economic and political contexts. Moreover, energy systems are highly interconnected with other parts of the full social-ecological system. Systems and complexity theory offer useful approaches to address energy futures. Tools such as systems and scenario modelling offer promising alternatives to isolated, sectoral planning methods. The information presented in this chapter provides a snapshot of some information that should be considered in moving towards an integrated energy systems approach to planning for a low carbon future in the Galapagos.

The literature and precedents examined above underscore the importance of taking an integrated view of achieving an appropriate energy mix. In particular, it is clear that technology is necessary, but not sufficient, in achieving energy security and sustainability. Human behaviour is one of the most important system components to consider. Individual and societal change must be as much a part of an energy transition as the exploration of innovative means of energy capture, storage and distribution.

There are opportunities to work with residential and small-scale operations on the Galapagos to implement solar hot water and photovoltaic systems. However, the greatest impact is to be made by focusing on tourists and the providers of tourism services. Visitors to the Galapagos Islands should understand the impact of their activity with respect to energy use. The author recently participated in a multi-day, marine-based ecotour. Although the guides were very knowledgeable about wildlife disturbance, introduced species and biological conservation in general, there was no mention of the energy challenges facing the islands. Guides should be required to include some educational content regarding energy on each tour. This could supplement the information that visitors receive when they arrive at the world's first 'green' airport on Baltra Island; a facility that is 100% is powered by renewable sources (ECOGAL 2016). Knowledge of the need for an energy transition may lead to greater philanthropic support from visitors to the Galapagos (Powell and Ham 2008). Ardoina et al. (2016) interviewed Galapagos tourists and reported a "connection to nature' effect, combined with newly acquired environmental knowledge, contributes to visitors' expressed intentions to donate in support of conservation of the Galapagos environment" (p. 11). These philanthropic intentions around wildlife conservation might be expanded to include energy issues.

Acknowledgements This chapter was influenced by the content of a panel presentation with participation from Byron Chiliguinga-Mazon (OLADE), Horacio Cuevas (Inter-American Development Bank), Sarah Jordaan (University of Calgary) and Adriana Valencia (Inter-American Development Bank). The Institute for Environmental Sustainability at Mount Royal University provided financial support to convene the panel.

References

Ardoina NM, Wheatona M, Huntb CA, Schuha JS, Durhamc WH (2016) Post-trip philanthropic intentions of nature-based tourists in Galapagos. J Ecot. https://doi.org/10.1080/14724049.2016.1142555. Accessed 20 March 2016

Bale CS, Varga L, Foxon TJ (2015) Energy and complexity: new ways forward. Appl Energy 138:150–159

Bueno C, Carta JA (2006) Wind powered umped hydro storage systems, a means of increasing the penetration of renewable energy in the Canary Islands. Renew Sust Energ Rev 10:312–340

Conservation International and WWF (2011) Adapting to climate change in the Galapagos Islands. Conservation International, Quito

Deniz C, Zincir B (2016) Environmental and economical assessment of alternative marine fuels. J Clean Prod 113:438–449

E-8 (2008) The San Cristobal wind and solar projects: displacing diesel-powered generation by renewable energy in the Galapagos Islands. In: E-8, Puertro Baquerizo Moreno. Galapagos, Ecuador, Isla San Cristobal

ECOGAL (2016) Galapagos ecological airport. http://www.ecogal.aero/en/about-us. Accessed 20 March 2016

Edgar GJ, Kerrison L, Shepherd SA, Toral-Granda MV (2003) Impacts of the Jessica oil spill on intertidal and shallow subtidal plants and animals. Mar Pollut Bull 47:276–283

Godina R, Rodriguez EMG, Matias JCO, Catalão JPS (2015) Sustainable energy system of El Hierro Island. Renewable Energy Power Qual J 13:232. http://www.icrepq.com/icrepq'15/232-15-godina.pdf. Accessed 2 May 2016

González JA, Montes C, Rodríguez J, Tapia W (2008) Rethinking the Galapagos Islands as a complex social-ecological system: implications for conservation and management. Ecol Soc 13(2):13. http://www.ecologyandsociety.org/vol13/iss2/art13/. Accessed 2 May 2016

Government of Barbados (2016) National sustainable energy policy [Revised]. http://www.energy.gov.bb/web/national-sustainable-energy-policy. Accessed 29 April 2016

Government of Ecuador (2015) Ecuador's Intended Nationally Determined Contribution (INDC). Unofficial English translation. www4.unfccc.int/.../Ecuador/.../Ecuador%20INDC%2001-10-2015%20-. Accessed 2 May 2016

Indian and Northern Affairs Canada (2010) Impact evaluation of the ecoENERGY for Aboriginal and Northern Communities. Government of Canada, Ottawa

International Galapagos Tour Operators Association (IGOTA) (2016) Challenges facing the Galapagos Islands. http://www.igtoa.org/travel_guide/challenges. Accessed 15 Mar 2016

IRENA (2015) Renewable energy policy brief: Ecuador. International Renewable Energy Agency, Abu Dhabi

Jácome C (2007) Energy subsidies in Galapagos. Galapagos Report 2006–2007. GNPS, GCREG, CDF and GC, Puerto Ayora, Galapagos, Ecuador. http://www.galapagos.org/wp-content/uploads/2012/04/socio10-energy-subsidies.pdf. Accessed 2 May 2016

Lougheed LW, Edgar GJ, Snell HL (2002) Biological impacts of the Jessica Oil Spill on the Galápagos environment. Final Report Version No.1.10. Charles Darwin Foundation, Puerto Ayora, Galápagos, Ecuador

Natural Resources Canada (2011) Status of remote/off-grid communities in Canada. http://www.nrcan.gc.ca/energy/publications/sciences-technology/renewable/smart-grid/11916. Accessed 3 May 2016

NCC (2016) Solar microgrids. http://www.nccsolar.com/about/. Accessed 3 May 2016

Pikaia Lodge (2016) The environmental sustainability of Pikaia Lodge in a land based tourism model of the Galapagos Islands. http://www.pikaialodgegalapagos.com/sustainability/the-environmental-sustainability-of-pikaia-lodge-in-the-new-land-based-tourism-model-of-the-galapagos-islands. Accessed 15 Mar 2016

Powell R, Ham S (2008) Can ecotourism interpretation really lead to pro-conservation knowledge, attitudes and behaviour? Evidence from the Galapagos Islands. J Sustain Tour 16(4):467–489

Sanchez AB (2014) Sustainability in the global tourism industry: good practice initiatives from the private and public sector. El Hierro 100% renewable energies. In: Egger R, Maurer C (eds) SCONTOUR 2014 - tourism research perspectives: proceedings of the international student conference in tourism research. Books on Demand, Norderstedt

Sevilla R (2008) An inconvenient truth and some uncomfortable decisions concerning tourism in the Galapagos. Galapagos Res 65:26–29

Shapiro A, Summers R (2015) The evolution of water management in Alberta, Canada: the influence of global management paradigms and path dependency. Int J Water Resour Develop 31(4):732–749

Simpson MCD, Scott M, Harrison N, Silver E, O'Keeffe S, Harrison M, Taylor M et al (2010) Quantification and magnitude of losses and damages resulting from the impacts of climate change: modelling the transformational impacts and costs of sea level rise in the Caribbean. United Nations Development Programme, Barbados

Tourtellot J (2015) Galápagos tourism backfires. National Geographic. http://voices.nationalgeographic.com/2015/01/05/galapagos-tourism-backfires/. Accessed 15 Mar 2016

Unruh GC (2000) Understanding carbon lock-in. Energy Policy 28:817–830

Westerman A (2012) An analysis of energy consumption on the Galapagos Islands: drivers of and solutions to reducing residents' energy consumption. J Public Int Aff 2012:109–130

Yu H, Martín MG, Jesús Sánchez M, Solana P (2015) Operational issues for the hybrid wind-diesel systems: lessons learnt from the San Cristobal Wind Project. In: Cortés P et al (eds) Enhancing synergies in a collaborative environment, lecture notes in management and industrial engineering. Springer, Cham

Chapter 3
Climate Change Policy as a Catalyst for Sustainable Energy Practice: Examples from Mainland Ecuador and the Galapagos

Irene M. Herremans and Mary-Ellen Tyler

Context

In 2011, Ecuador's National Electricity Board (CONELEC) identified Ecuador's electricity generation sources as: "Hydroelectric: 60.18%, Thermoelectric: 31.94%, Other renewable sources: 0.82%, Imported electricity: 7.05%" (Villa Romero 2013). While hydroelectricity has a relatively low carbon footprint, it can also be susceptible to regional droughts, rainfall variability, and climate change impacts affecting rainfall amounts and seasonal precipitation patterns. For example, Ecuador experienced unusually low rainfall in 2011, which resulted in low reservoir levels which in turn resulted in rolling electricity outages affecting over one million people in in different quadrants of the capital city of Quito. Similarly, the amount of glacial meltwater sources available to Empresa Eléctrica Quito decreased by 50% between 1978 and 2008 (Villa Romero 2013).

Through "Decree 1815" Ecuador established a National Strategy for Climate Change to identify climate change actions for the period 2012–2025 (Nachmany et al. 2015, p. 3). Ecuador's related National Plan for Good Living (2013–2017) (Republic of Ecuador 2013) includes strategies for climate change adaptation and mitigation in partnership with Ecuador's National Environmental Policy. Objective 10 of this National Plan promotes diversification of energy mix within the context of the National Climate Change Strategy (Nachmany et al. 2015, p. 5). The intention is "to reduce net emissions through increased efficiency in production of electricity"

I.M. Herremans (✉)
Haskayne School of Business, University of Calgary, Calgary, AB, Canada
e-mail: irene.herremans@haskayne.ucalgary.ca

M.-E. Tyler
Faculty of Environmental Design, University of Calgary, Calgary, AB, Canada
e-mail: tyler@ucalgary.ca

through the development of renewable energies (Nachmany et al. 2015). In an effort to diversify the electricity generation grid mix with more sources of renewable energy, CONELEC initiated Regulation 004/11 in 2011, which is a feed-in tariff to support the development of "solar photovoltaic, wind, geothermal, biomass, biogas, and hydro-energy" (Nachmany et al. 2015). This competitive concession contracting was intended to secure renewable electricity production from alternative sources using 20-year government contracts as an incentive for development.

Case One: A Radical Approach to Solar Energy Development

In 2010, Radical Energy, a Canadian solar photovoltaic company specializing in the development of solar projects from the idea stage to the operational stage, undertook development planning for two large-scale solar power projects in Ecuador. These two projects (Cóndor Solar and Solarconnection) had a combined planned capacity of 63.5 MWp/3 MWac. Although these projects were ultimately not constructed, their history provides important lessons useful for other locations interested in developing sustainable energy projects. Specifically, Radical Energy's planning and development approach was based on three fundamental principles:

- Produce ethical returns for investors.
- Focus on environment, health, and sustainability.
- Ensure maximum social benefits and knowledge transfer.

The initial land area planned for both projects was 147 hectares (ha) which is roughly equivalent in size to 133 soccer fields. The area chosen was dry with poor potential for agricultural production. According to software calculations using RETScreen and PVSyst, the annual electricity generation would be 106 GWh (196,000 MWh), which was equivalent to an annual electrical load of approximately 100,000 households in northern Ecuador.

Project planning included approximately 211,000 three hundred watt solar panels connected to a 69 kV distribution line bisecting the project area. As planned, the two projects would offset an estimated 47,000 tonnes of CO_2 annually based on the 2011 carbon intensity of the Ecuadorian grid. This would be equivalent to removing 8,700 mid-sized cars from the road or 4,392 hectares of forest absorbing CO_2 annually. Projected employment opportunities generated by Cóndor Solar and Solarconnection development included 500–700 jobs over a 7–10-month construction period and 15–25 full-time positions post-construction related to site security, panel washing, and general systems monitoring and maintenance.

The total capital expenditure estimated for project construction was $212 million in United States currency. In order to fund project construction, an international financial consortium was created which involved companies from Spain

(Solarpack Technologies, SENER Engineering), the United States (Sunwize Technologies), and Canada (Radical Energy, Solexica Development Corporation, JCM Capital). Approximately $8.5 million was initially raised to cover land purchases, legal expenses, government performance bonds, operational costs, and bank due diligence costs. Three international banks were involved in project financial planning including the Inter-American Development Bank, Proparco (a private arm of the French Development Bank), and the FMO (a private arm of the Dutch Development Bank).

Community and environmental investments of $200,000 annually for the duration of the 20-year concession contract were included in the project development plan. With an estimated project life of 35 years, this would produce $6 million in the first 20 years and considerably more once project ownership transferred to the government of Ecuador after year 20. These monies would be created by electricity revenues and intended to support education through scholarships and small business development through micro-finance loans, as well as other environmental conservation and direct investment community projects. Although these community and environmental monies were never realized because the projects were not constructed, during the planning phase, the development consortium did donate $40,000 to an organic farming NGO working adjacent to the planned projects.

Solar Project Development Phases

The role of Radical Energy included project originator, planner, and developer. Radical investigated, planned, and developed project plans to the point where the projects were able to attract financing, engineering, procurement, and construction partners and ultimately project buyers. The scope of services and activities handled by Radical Energy was multifaceted and required a very broad understanding of all aspects of energy project development including legal, policy, environmental regulations, and stakeholder relations in addition to engineering, finance, and construction. Radical Energy's phases of solar project development planning are identified below, but in practice they frequently overlapped and were iterative because of feedback from various phases:

- Market assessment and research
- Project site investigation and development planning
- Technology and equipment investigation and procurement
- Partnership development and financing
- Procurement, construction contracting, and labor force recruiting
- Project ownership transition (on-going operations and maintenance logistics)

Market Assessment and Research

Radical Energy's activities in this phase focused on two key activities: pre-feasibility analyses of a project's specific financial, regulatory, and legal requirements, and identification of country-specific risks and strategies to avoid, transfer, or mitigate for investors and developers. As the project development planner, Radical Energy analyzed several sources of solar data utilizing global horizontal irradiance as the key indicator. Using these results, the areas with the best solar irradiation were cross-referenced with the location of electrical distribution and transmission lines. Results were then used with RETScreen software to run a financial analysis of the project's siting and estimated annual electrical generation.

Project Site Investigation and Development Planning

This phase involved the following major activities: investigate alternative development sites, site selection, negotiate a land control contract, initiate contact and plan for community and social engagement, and conduct social and environmental impact scoping. Although government information sources were used to map the location of distribution and transmission lines, Radical Energy's on-the-ground Geographic Positioning System (GPS) cross-checks found this mapping information to be inaccurate by several kilometers. Because of the significant challenge of finding a large enough area to accommodate both projects, Radical Energy personnel used a local taxi truck fitted with a camera and GPS to find potential sites and confirm the locations of electrical lines. Finally, property of sufficient size for a 69 kV distribution line running through it was identified as having sufficient capacity to transport future solar electricity production.

Radical Energy executives had learned one very important aspect of sustainable energy project planning while participating in the Sustainable Energy Development (SEDV) Master of Science Program delivered cooperatively by the University of Calgary and the Universidad San Francisco de Quito in Quito, Ecuador. This was the importance of stakeholder and community engagement in project development. Acquiring stakeholder buy-in and support early in the project planning process was crucial to moving forward with the projects. Radical began engaging with almost all stakeholders 2 years before planned construction, and the company used the International Finance Corporation's (IFC) Performance Standards on Environmental and Social Sustainability (updated from 2006) to plan and manage interactions with the local communities and for the environmental impact assessment of the project (IFC 2012). This process created strong relationships with stakeholders including numerous government departments; formal and informal consultations with aboriginal communities; and women's groups, local community councils, and the neighboring organic farming NGO. An Inter-American Development Bank (IDP) specialist who was assigned to review the project work commented on it as being one of the best engagement processes they had reviewed in Latin America (Dick 2014).

Technology, Equipment Investigation, and Procurement

The engineering analysis in this phase involved several steps: grid connection viability, technology and equipment selection, procurement, and systems engineering. The purpose of grid connection viability was to determine if the Cóndor Solar and Solarconnection projects could connect with existing transmission and distribution lines and transport the electricity into the local grid. Activities included analysis of transmission and distribution interconnection requirements, a preliminary grid capacity analysis, and negotiations for solar project interconnections with the local electrical grid infrastructure.

Based on the results of this analysis, Radical Energy and its partners then investigated suitable solar PV technology options for the project site. With a consideration for cost-effectiveness, activities included consultation with engineering procurement and construction firms; negotiation, audit, signing of purchase contracts; and finalizing selection of technology and engineering firms for construction and installation. System engineering activities included utilizing expertise from solar engineers to design the final electrical schematics and layout for both projects and obtain final approvals from government authorities.

Partnership Development and Financing

Radical Energy investigated suitable partners to share both the operational and financial risk for project development. Activities involved strategic partner identification, which included preparing project portfolios, audits, and negotiating investment contracts with the assistance of legal experts. Lender identification required developing project marketing, financial prospectus material, and specific information for bank officials. Equity investors identification involved working through due diligence reviews of financial and project information with private equity investors.

Procurement, Labor Force Recruitment, and Construction Contracting

Before necessary equipment and materials can be purchased and construction can begin, all applicable laws and policies must be met. All legal documentation required must be completed for project approval including local permitting and zoning requirements. Establishing and maintaining strong government connections were critical in working through all the requirements for final project approvals prior to proceeding with procurement and contracting. Specifically, two onsite archeological studies, a forestry license to move harvested trees offsite, solar interconnection approvals with the local electricity grid distribution company, a social environmental assessment study, full project design approvals, and construction approvals were all required prior to project construction contracting.

Project Ownership Transition (On-Going Operations and Maintenance Logistics)

Upon completion, project assets were expected to be transferred to an asset holder. Planning for this transfer requires ensuring the project is functioning as an autonomous electricity generating plant and fully operational under a local operations and maintenance company.

Solar Development Lessons Learned

The Cóndor Solar and Solarconnection projects were not completed for political reasons beyond Radical Energy's control. However, the experience with the planning for these two solar projects provided some important lessons specifically related to government policy and community and stakeholder engagement.

Government policy and related incentives while intended to support solar energy development also add a significant degree of risk and uncertainty to infrastructure projects like Radical Energy's two solar projects. Such risks are difficult to mitigate in the planning and development stage, and it is not uncommon for government policy to change in the middle of development, which in turn, completely changes the economic viability of a project. Because all the phases of solar project development as outlined above can take many years to fully implement, it is critical that policy in force at the time of project planning be grandfathered to enable project completion. This situation is not unique to Ecuador or developing countries. For example, both Spain and the provincial government of Ontario in Canada have had policy changes that caused not only projects but a fledgling solar industry sector to falter or fail to reach full potential (Fraser Institute 2012; McKitrick and Adams 2014). In order for government policy to support the growth of and transition to renewable energy, "TLC" (transparency, longevity, and certainty) is required (Deutsche Bank 2009, p. 4). Sustainable energy development projects require years of planning work and millions of dollars in investment and legal agreements before construction can begin and power generated. By creating long-term, transparent policy and regulations to support this investment of time and money, investors and the public can be confident that investment dollars can be repaid from long-term electricity sales revenues.

Engagement with stakeholders, especially with communities geographically located near the project is a valuable social learning experience at all stages of project development. Honesty, transparency, and relationships built on trust over time are essential throughout the entire project engagement process. However, there is a risk that some project information may receive a negative response from stakeholders that could jeopardize the project. As well, project personnel may not always be pleased with what they hear in the engagement process, but concerns and motivations of the community will help them to understand their own project better. Ultimately, the objective is to get the project implemented and address stakeholders' concerns early in project development. All types of engagement processes, from large town hall meetings to smaller meetings with community leaders, small group discussion groups, informal votes, and prioritization of alternatives, can occur during the stakeholder engagement process. Although input can range from useful, well-thought-out comments to angry off-topic attacks, it is all part of the learning process. Community meetings help to convey how company investments might be allocated to create a sense of trust between project developers and the community. An open dialogue can assist in forging a bond that can help a project get through some difficult times. Initially, it would seem that a large-scale solar project would be an easier sell to a community than a large-scale open pit mine or even a wind power project. But, based on Radical Energy's project development experience, in any type of project, stakeholders want to be treated as if they are important and part of the process (Dick 2017). Stakeholder influence and opposition can cancel a project, cause delays, or change government policy. Therefore, although community and stakeholder engagement takes time and money, it is fundamental to successful sustainable energy project development.

Acknowledgements Information specific to Radical Energy project was generously provided by Ryan Dick, Terrapin Geothermics, Edmonton, AB, Canada, who directed Radical Energy's project work in Ecuador including the project in this case and also participated in the 2014 World Summit in the Galapagos.

Case Two: Adding Energy Efficiency into the Energy Mix

In 2016, graduate students from the Sustainable Energy Development Master of Science degree program at the University of Calgary undertook internships in the Galapagos. They worked on a project in the community of Puerto Ayora and the Island of Santa Cruz initially started by the World Wildlife Fund and the Republic of Ecuador's Ministry of Tourism in the Galapagos. They undertook energy and water audits of a sample of tourist hotels to help identify where energy efficiency could be improved to better manage growing energy demands from increasing land-based tourism pressures. The results of these audits which included interviews with hotel managers and staff and lessons learned from these results are described as follows.

Energy Audit Results

Figures 3.1 and 3.2 compare the audit results for three medium hotels (MSH) and three small hostels (SSH). Results have been separated into high and low tourist season estimates and low-, medium- and high-energy consumption categories using

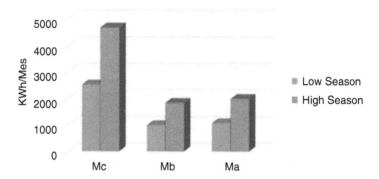

Fig. 3.1 Energy audit comparisons of medium-sized and small-sized hotels

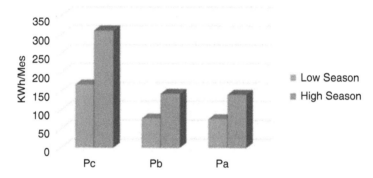

Fig. 3.2 Comparison of small-sized hotels energy use in high and low tourist seasons

median values from the range of audit results. The total kWh/month for the MSH far exceeds the SSH total because of the greater number of energy services provided by the MSH such as air conditioning, mini-refrigerators, and larger buildings with significantly more lighting in hallways, staircases, lounges, and lobbies, which all contribute to higher electricity consumption.

The two outliers in each figure (Pc and Mc) are the result of their higher electricity consumption due to air conditioning. Specifically, Pcs offered air conditioning in more rooms than the other hostels. Mc's high electricity use was due to the type of air conditioners used. The air conditioner units in Mc were rated at 3500 W in all the guest rooms that were assessed. This is a significantly higher power rating for new AC units than observed in other hotels and rated at 1100–1200 W. In examining the three major electricity consuming devices in Fig. 3.3, AC consumed 43%, ceiling lighting 41%, and refrigeration (including freezers) 16% of the total electricity consumed by all six hotels combined.

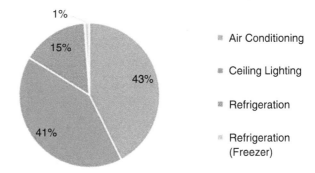

Fig. 3.3 Total electricity use by device type

Lighting

From the overall energy consumption audit of the six assessed hotels, ceiling lighting contributed 41% of the energy use per year. If all the light bulbs in the six hotels were replaced by light emitting diode (LED) bulbs, lower energy consumption could be achieved. Only one of the six hotels has switched from compact fluorescent light (CFL) and incandescent to LED for communal areas. Table 3.1 compares the three lighting types used in the hotels audited. The specifications for the light bulbs in Table 3.1 are based on retail prices and bulb types available in hardware stores in Puerto Ayora and Santa Cruz, and prices are in US Dollars.

The six hotels use less than 10% incandescent light bulbs. The transition from incandescent lights to CFL bulbs resulted from a previous best practices program implemented by the World Wildlife Fund (WWF) and Ecuador's Ministry of Tourism (WWF Program Galapagos 2013) to promote ecotourism on the Islands. A similar program could phase out CFL use, as LED bulbs use approximately 50% less energy. Given the current energy mix in the Galapagos is approximately 70% diesel, greenhouse gas (GHG) reduction and energy conservation are important (Dove 2014).

Potential LED Savings

Table 3.2 illustrates the range of savings possible if all six of the sample hotels transitioned to using LED light bulbs. The assumptions used in generating Table 3.2 include the following: lights are on in hotel communal areas 4 h/day; lights are on in guest rooms 3 h/day; there is an emission factor of 0.96 kg CO_2/kWh which would require 13 kg CO_2/3-year to offset emissions; and energy mix is assumed constant for the lifetime of the LED bulbs.

As illustrated in Table 3.2, the potential savings per year of LED bulbs is significant. A medium hotel could save an estimated US$400 to US$530 per year and a small hostel an estimated US$70 to US$130 per year with the payback period ranging from 1 to 2 years.

Table 3.1 Comparison of three types of hotel light bulbs

Per bulb	LED	CFL	Incandescent
Power	8.5 W	20 W	40 W
Price	$7~$10	$4–$6	$2~$4
Life-span	25,000–50,000 h	8000–10,000 h	1200–3000 h
Environmental impact	–	Contains mercury	Creates more waste

Table 3.2 Range of potential savings from LED light bulbs

Potential savings	Combined total
Utility money saved per year (USD)	1056
Money saved on maintenance/year (USD)	610
Total money saved/year (USD)	1666
Diesel saved/year (L)	3749
CO_2 emissions offset (ton/year)	7
CO_2 emissions offset in lifetime (ton)	130
Trees required to offset the emissions	551

Air Conditioning

Air conditioning (AC) was the highest energy-consuming device in the hotel audit using 43% of energy consumed. Most hotels in Galápagos do not have building designs that incorporate shading, thermal mass, or natural ventilation. Hot season temperatures can extend from January into July (Charles Darwin Foundation 2016). Audit results found that fans were not considered as a viable alternative to AC in hotels. AC units used are relatively new (2013) and have different function modes. However, interviews with the hotel managers suggest these energy saving features are not being used. The International Energy Agency (IEA) suggests (IEA 2013, p. 3): "In hot climates, the energy savings potential from reduced energy needs for cooling are estimated at between 10% and 40%. More than 40% of the savings expected in heating and cooling energy demand under a low carbon scenario can be directly attributable to improvements in the building envelope."

To reduce electricity consumption rates and decrease tourist sector carbon footprints, both hotel owners and tourists need to embrace an energy conservation attitude. AC units need to be set at an appropriate level. In the Galápagos, AC can be preset in different modes during the year to work optimally in each season. There is also a fan function that can be used to reduces electricity consumption of AC units by 85%. Many international hotels encourage their customers to save water and energy through educational initiatives such as sustainability labeling, and this can be done in the Galapagos.

3 Climate Change Policy as a Catalyst for Sustainable Energy Practice: Examples...

Table 3.3 Potential savings from proper sealing of windows and doors

Number of assessed ACs	73
Average of energy consumed by ACs in assessed hotels (kWh/y)	99,808
Energy saved by proper sealing	20%
Average of energy saved in ACs at assessed hotels (kWh/y)	19,962
Potential US$ saved	1996

Table 3.4 Reduced energy use from fans

Number of assessed ACs	73
Average of energy used for ACs in assessed hotels (kWh/y)	99,808
Energy reduced by using a fan instead of an AC	85%
Average of energy saved in ACs at assessed hotels (kWh/y)	16,967
Potential US$ saved	1,697

Improving Sealing

The Energy Star program (2009) states the properly sealed rooms could save could save almost 20% in energy costs. Table 3.3 illustrates the potential energy savings from proper sealing of hotel windows and doors for the six hotels audited which could provide a potential saving of 19,962 kWh/y which is equivalent to US$1996 in savings.

The Rainforest Alliance (n.d.) has suggested that fan use represents only 15% of AC energy requirements. The increased use of fans instead of AC units could reduce hotel energy use as illustrated in Table 3.4, which indicates that fans could reduce AC energy consumption by 20% on an annual basis.

If these assumptions and numbers can be extrapolated to include all the hotels in the Galapagos, then the potential reduction in energy use could have a significant impact. Increasing the efficiency of energy use is an important component of a sustainable energy mix transition in the Galapagos.

The Water-Energy Nexus

Water and energy use are connected. Energy is required for the collection, treatment, distribution, pumping, and heating of potable water. Energy consumption related to water use ranges from 4% to 19% of electricity consumption (Copeland 2014). Energy use related to water use in the Galapagos has not been calculated but is an overlooked aspect of energy management on the Islands. Increasing population growth and land-based tourism has increased water use significantly and specifically in Santa Cruz. The municipal water supply in Santa Cruz is currently at capacity.

Water Audit Results

The same six hotels involved in the energy audit process were also involved in a water use audit. The water audit included the following: bathroom faucets, showerheads, toilets, laundry facilities, kitchen facilities, outdoor landscape irrigation purposes, housekeeping and maintenance, pool, and spa facilities. Audit results identified the areas with the highest water saving potential. Strategies for water savings were provided to each hotel, and a small report and discussion held with hotel staff on implementing the strategies for water savings, which can increase the energy efficiency of hotel operations. Water-saving opportunities and estimated savings are projected in Table 3.5. It should be noted that water consumption in guest bathrooms is very dependent on the occupancy rate and the water consumption behavior of guests. For the purposes of Table 3.5, high season is assumed to be from December to July at 65% occupancy, and low season is assumed to be from August to November at 35% occupancy. The towel and linen reuse assumes 10% reuse by guests. Water consumption in guest bathrooms is very dependent on the occupancy rate and the water consumption behavior of guests. As illustrated in Table 3.5, the potential water savings in small hotels translates into approximately the volume of one to two domestic swimming pools. Water saving in medium hotels translates to roughly the volume of three to seven domestic swimming pools.

Table 3.5 Water-saving opportunities

Low-cost water-saving actions	High efficiency faucet aerators	Low-flow showerheads	Repairing leaking toilets	Towel and linen reuse program	
Approximate cost across hotel (US$)	5–10 per faucet	15–75 per shower	50–125 per toilet	Cost of printing signage	Total estimated water-saving opportunities
Annual water savings in a small hotel (liters)	4,700–12,000	34,000–64,000	7,000–14,000	5,400–7,300	51,000–97,000
Annual water savings in a medium hotel (liters)	25,000–145,000	48,000–96,000	100,000–200,000	23,000–32,000	196,000–473,000

Desalination Water Treatment Plant

In 2016, a reverse osmosis desalination plant became operational on the Galapagos Island of Santa Cruz. The desalination plant produces safe water low in chloride content that is distributed throughout the Island. However, a desalinization plant has significant costs and energy demands (Reyes et al. 2015). The demand for desalinization plant water is highest for residential and hotel use (Reyes et al. 2015). The desalination process requires significantly more energy than other water treatment methods. The process generally requires three steps: (1) the pretreatment of the reverse osmosis membranes, (2) the reverse osmosis process, and (3) the posttreatment of the permeable membrane. The highest energy consumption is usually in the reverse osmosis step (WRA 2011).

Increasing Efficiency for the Energy-Water Nexus

Behavioral change is key to lowering energy and water consumption. Creating greater awareness about the importance of energy efficiency and the water-energy nexus for sustainable energy mix in fragile environments is important in dealing with population growth and increasing tourism pressure in the Galapagos. Table 3.6 identifies specific actions to improve performance. Each hotel audited was provided with specific recommendations customized to their performance and needs.

Table 3.6 Specific actions to improve energy and energy-water nexus efficiency

• General suggestions	• Train employees to conserve energy and water
	• Induce behavioral changes in the tourists
	• Informative and accurate signage for "save energy and water"
	• Automatic key cards, external switches
• Air conditioning	• Clean filters
	• Proper insulation
	• Promote use of fan feature in ACs
• Refrigeration	• Proper sealing on door
	• Located in cool place, distance from the wall
	• Discourage empty refrigerators
• Lighting	• Encourage use of LED bulbs
	• Use of clean bulbs
	• Use of reflectors, open lamps to improve illumination
	• Use of external lights
• Water conservation	Install low-flow toilets and showerheads
	Repair leaking toilets
	Install aerators in faucets
	Utilize a towel and linen reuse program

Lessons Learned

The results of the energy and water audits from this small sample of hotels in the Galapagos Islands suggest that low-cost efficiency upgrades and minor maintenance of devices could improve energy efficiency and in the process save energy and money. Energy efficiency is necessary for the Galapagos Islands to transition to a sustainable energy mix. Energy efficiency that reduces energy demand makes this transition more feasible. There are numerous energy efficiency opportunities in the areas of AC, refrigeration, and lighting which can contribute to cumulative energy savings for Galapagos. Creating shared value among institutions, businesses, and consumers is pivotal to implementing these measures. Audit results suggest there is a lack of knowledge and tools available to businesses for implementing efficiency measures. Opportunities exist for educational initiatives to fill this gap. Additional initiatives can be directed at influencing tourist behavior. Together these initiatives have the potential to advance sustainable energy use.

Acknowledgements Information specific to energy and water audits described in this case was generously provided by Alonso Alegre, Connor Bedard, Kasondra Harbottle and Namrata Sheth who conducted the audit project as participants in the Galapagos internship summer program of the Master of Science in Sustainable Energy Development degree program, University of Calgary, Calgary, AB, Canada

Conclusions

Both the solar energy project case and the energy efficiency audit project demonstrate a variety of lessons learned and critical factors influencing sustainable energy mix development. Both the production of new renewable energy generating capacity and improving the efficiency of existing energy use are components of developing sustainable energy mix options in specific circumstances. While technological innovation plays a key role in developing sustainable energy, both of the cases reviewed here illustrate the importance of policy innovation and social learning through engagement and behavior change as having an equally key role in sustainable energy mix development.

References

Charles Darwin Foundation (2016) Meteorological database. http://www.darwinfoundation.org/datazone/climate/

Deutsche Bank (2009) Paying for renewable energy: TLC at the right price. DB Climate Change Advisors, Deutsche Bank Group, New York. https://institutional.deutscheam.com/content/_media/1196_Paying_for_Renewable_Energy_TLC_at_the_Right_Price.pdf

Dick R (2014) Radical Energy CEO, personal communication

Dick R (2017) Radical Energy CEO, personal communication

Dove S (2014) Sustainable energy in the Galapagos. Bridges 8. Retrieved from: https://www.coastal.edu/media/academics/bridges/pdf/Dove_Final.pdf_ab.pdf

Energy Star (2009) A guide to energy-efficient heating and cooling. Retrieved from: www.energystar.gov/ia/partners/publications/pubdocs/HeatingCoolingGuide%20FINAL_9-4-09.pdf

Fraser Institute (2012) Ontario households to pay extra $285 million annually for electricity. https://www.fraserinstitute.org/article/folly-ontarios-renewable-energy-program-provides-warning-other-governments-ontario. Accessed 17 Mar 2016

International Finance Corporation (IFC) (2012) IFC performance standards on environmental and social sustainability. Washington, DC. (www.ifc.org/sustainability) https://www.ifc.org/wps/wcm/connect/c8f524004a73daeca09afdf998895a12/IFC_Performance_Standards.pdf?MOD=AJPERES

McKitrick R, Adams T (2014) How green energy is fleecing Ontario electricity consumers, Special to Financial Post. FP Comment. http://business.financialpost.com/opinion/how-green-energy-is-fleecing-ontario-electricity-consumers. Accessed 16 Mar 2017

Nachmany M, Frankhauser S, Davidova J, Kingsmill N, Landesman T, Roppongi H, Shleifer P, Setzer J, Sharman A, Stolle Singleton C, Sundaresan J, Townshend T (2015) Climate change legislation in ecuador. an excerpt from the 2015 global climate legislation study a review of climate change legislation in 99 countries. http://www.lse.ac.uk/Granthaminstitute/wp-content/uploads/2015/05/ECUADOR.pdf. Accessed 16 Dec 2016

Rainforest Alliance (n.d.) Buenas Prácticas para Turismo Sostenible. Retrieved from: www.rainforest-alliance.org/business/tourism/documents/tourism_practices_guide_spanish.pdf

Republic of Ecuador (2013) National development plan/national plan for good living, 2013–2017. Summarized Version. National Secretariat of Planning and Development (www.planificacion.gob.ec), Quito, Ecuador. Accessed at http://www.planificacion.gob.ec/wp-content/uploads/downloads/2013/12/Buen-Vivir-ingles-web-final-completo.pdf

Reyes M, Trifunovic N, Sharma S, Kennedy M (2015) Water supply assessment on Santa Cruz Island: a technical overview of provision and estimation of water demand. In: Galapagos report 2013–2014. GNPD, GCREG, CDF and GC, Puerto Ayora, Galapago, pp 46–53

Villa Romero JF (2013) http://latinamericanscience.org/2013/11/ecuador-steps-up-its-fight-against-climate-change/. Accessed 16 Dec 2016

World Wildlife Fund (2013) Galapagos stories. WF Program Galapagos, https://www.worldwildlife.org/stories?place_id=the-galapago

Water Reuse Association (WRA) (2011) Seawater desalination power consumption. WateReuse Desalination Committee's White Paper. watereuse.org/wp-content/uploads/2015/10/Power_consumption_white_paper.pd

Chapter 4
Biofuels in the Energy Mix of the Galapagos Islands

Irene M. Herremans and Arturo Mariño Echegaray

Introduction

The most popular method of transportation for visiting the Galapagos Islands is still via cruise ship. However, island tourism is moving inland. As many cruise ships are not locally owned (Quiroga 2014) or make few stops in the populated areas, inland tourism creates new economic and social benefits for the inhabited islands of Santa Cruz, San Cristobal, Isabella, Floreana and Baltra. According to 2015 census results, a total of 25,244 people inhabit the Galapagos Islands. Santa Cruz Island had 15,701 inhabitants or approximately 62% of the Galapagos population. San Cristobal and Isabela had 29% and 9%, respectively. The average annual growth rate for the Galapagos Islands for the 5 years of 2010–2015 was 1.8% (Instituto Nacional de Estadística y Censos – Gobierno Nacional de la Repùblica del Ecuador 2015).

Although there are economic and social benefits associated with land-based tourism, the potential also exists for environmental impacts and cumulative effects. Expanding inland tourism is likely to increase the need for transportation related to transport for tourists and related goods and services. Fuel and electricity demands are likely to increase with increasing development of land-based tourism related to increasing land-based transportation as well as for boats for island excursions and related recreation including scuba diving and snorkelling. Historically, electrical power in the Galapagos has been provided by generators using diesel fuel shipped from the mainland. To meet increasing electrical demand from the development of new land-based tourism, additional fuel for diesel power generators will need to be transported from the mainland in order to keep both tourists and local residents

I.M. Herremans (✉)
Haskayne School of Business, University of Calgary, Calgary, AB, Canada
e-mail: irene.herremans@haskayne.ucalgary.ca

A.M. Echegaray
Faculty of Environmental Design PhD Program, University of Calgary, Calgary, AB, Canada

© Springer International Publishing AG 2018
M.-E. Tyler (ed.), *Sustainable Energy Mix in Fragile Environments*,
Social and Ecological Interactions in the Galapagos Islands,
https://doi.org/10.1007/978-3-319-69399-6_4

comfortable with air conditioning, lights, water and appliances. However, the increased shipping of diesel fuel brings with it greater potential for risk to the marine environment. Tanker accidents and spills have previously occurred in the Galapagos with negative effects on the quality of the protected marine environment. For example, in 2001, the tanker *Jessica* ran aground in the bay of Puerto Baquerizo Moreno off the island of San Cristobal. The tanker was carrying 160,000 gallons of diesel fuel for delivery to Baltra Island and 80,000 gallons of bunker fuel for the tourist boat *Galapagos Explorer*. Due to favourable wind conditions at the time of the spill, the oil slick dispersed widely and away from land. However, subsequent research showed the death rate of marine iguanas increased due to the spill's effects on their ocean food supply (Lougheed et al. 2002). Meeting the annual needs of an estimated 200,000 land-based tourists in addition to local residents will not be without increased risk of marine spills. The marine and land spill risk is particularly serious because the World Heritage designation given to the Galapagos Islands in 2007 is a primary driver for tourism.

Sustainable Energy Mix Development

Ecuador's policy of "Zero Fossil Fuels on the Galapagos Islands" (Ministerio de Electricidad y Energía Renovable Sub-Secretaría de Energía Renovable y Eficiencia Energética 2016) provides a framework for exploring alternative energy mix options to meet growing energy needs associated with land-based tourism, local economic development and local population growth while managing the marine and terrestrial ecological integrity of the Islands as a World Heritage site. The original timeframe for policy implementation and shifting energy mix targets to renewable energy sources was 2020. Potential sources for creating a zero fossil fuels energy mix include wind, solar, biofuels and possibly marine energy. However, any selected source will have to demonstrate there is no risk to critical species and critical marine and terrestrial habitats. In a World Heritage Site characterized by fragile marine and terrestrial environments, integrating the economic, social, and environmental dimensions of sustainable energy is critical to support the entire socioecological system of the Galapagos Islands.

Wind and solar are commonly known renewable energy sources. But, there is also potential for plant-based biofuels to serve as an alternative to diesel and bunker fuel. This plant-based alternative has the potential to lower the impact of spills in both marine- and land-based environments. As illustrated in Fig. 4.1, the 2012 energy mix in the Galapagos included 77.72% diesel (fossil fuel), 20.82% wind, 1.20% biofuel and 0.27% photovoltaic energy sources. However, Ecuador's 2017 policy target for a sustainable energy mix by 2020 identified a shift to 48.90% biofuel, 43.73% wind and 7.38% photovoltaic sources with no fossil fuel contribution (MEER 2012, p. 6).

Ecuador's policy to achieve 100% renewable energy use in the Galapagos by 2020 is one of the most ambitious renewable energy development initiatives in the

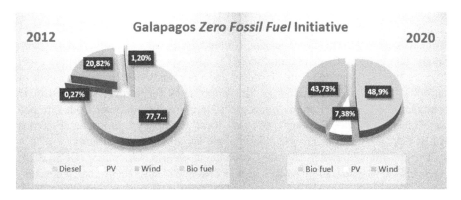

Fig. 4.1 Galapagos energy mix transition 2012–2020. Source: Adapted from Ministry of Electricity and Renewable Energy (MEER) (2012)

world (Gruber 2014). To date, finding an appropriate energy mix for the Galapagos Islands has involved establishing resource and expertise partnerships among Ecuador's government ministries and international organizations to explore the use of biofuels and develop pilot projects for testing. Specifically, from 2008 to 2011 Ecuador's Ministry of Electricity and Renewable Energy (MEER) and Ministry of Agriculture (MAGAP) worked with Germany's development agency Deutsche Gesellschaft für Internationale Zusammenarbeit (GIZ) on biofuel projects. The use of renewable energy is economically and technically feasible in the Galapagos. But, sustainable energy mix development requires capacity building through technical training for local people for project installation and on-going operations. The GIZ sponsored biofuels project on Floreana Island involved a permanent population of approximately 200 residents. The Floreana pilot study tested the use of biofuel from Jatropha for thermal electricity generation. The Ecuadorian province of Manabi provided the feedstock because of its plentiful supply of Jatropha curcas and need for rural economic development. Approximately 700 small farm families were involved in Jatropha collection for floreana. Through an agreement between MEER, Inter-American Institute for Cooperation on Agriculture (IICA), and GIZ, an extractor was provided to extract oil from Jatropha. Two rural cooperatives (COOPIÑOM and COOPROCERMA) were engaged as the operators of the Compac Tropha Oil Mill and transported the oil to Floreana Island. The international Climate Initiative Report – power generation using Jatropha oil on the Galapagos Islands published in January 2016 – shows that the Jatropha harvest expanded from 137 tonnes in 2011 to 213 tonnes in 2012, and Jatropha oil yields from 36,000 L (2011) to 50,000 L (2012) (Federal Ministry for the Environment, Nature Conservation, Building and Nuclear Safety, International Climate Initiative 2016).

The Instituto Nacional Autónomo de Investigaciones Agropecuarias (INIAP) provided support in selecting the plants with the best characteristics for oil production. Approximately 3000 families in Manabi benefitted production from oil. German experts worked with local people to improve the quality of the oil produced to meet German quality standard DIN-51605 (2016–01) (Strauss et al. 2011). Subsequently, an International Climate Initiative Report (Renewable Energy Resources for the Galapagos Islands) was published in September 2015 on the project's initial stages (Federal Ministry for the Environment, Nature Conservation, Building and Nuclear Safety, International Climate Initiative 2015). The report suggested replacing fossil fuels in power generators with biofuels and coordination of hybrid systems of Jatropha oil-operated generators as back-up systems for primary solar and wind powered energy production. Two electric generators using biofuel from Jatropha were installed on Floreana and a local company, Empresa Elèctrica (ELECGALAPAGOS), took over continuing operation of the generators.

The Instituto Interamericano de Cooperación para la Agricultura (IICA 2013) considers the Manabí Jatropha oil project an example of sustainable energy development. The project avoided the food versus fuel conflict because no agricultural land, or rainforest, or irrigation, or chemicals were used, and gender, social, and property rights were respected. For example, the two cooperatives involved (COOPIÑOM and COOPROCERMA) are working to earn the Roundtable for Sustainable Biomaterials (RSB) certification (RSB 2012), approved by the European Commission under the European Renewable Energy Directive. To receive this certification, all agricultural steps in the value chain such as "nut preparation, planting, hedge trimming and maintenance, harvest, storage and transport" must meet sustainable biomaterial production standards (IICA 2013). RSB certification will help make Manabi produced Jatropha oil more competitive on the world market and increase its market value.

Jatropha Oil Production

Jatropha curcas is a drought tolerant native Ecuadorian plant that can produce biodiesel. These native Jatropha plants can live to 50 years and grow up to a metre annually. Jatropha can grow in marginal soil conditions and in good soil conditions, yields can be as high as 12 tons per hectare (Carvajal 2012). Jatropha is not considered an invasive species in the Galapagos Islands and is not a threat to food security as the fruit is not edible (Carvajal 2012). The transformation of Jatropha into oil for biodiesel feedstock includes processing and packaging the oil, transporting the oil, and using the oil in biodiesel-compatible generators (MEER 2012). The production of Jatropha biodiesel is a chemical process. Oil molecules (triglycerides) are disconnected and reconnected to methanol molecules to form a methyl ester oil. Alkali is needed to catalyse this reaction, and glycerol is formed as a side product (FAO 2010). According to Food and Agriculture Organization of the United Nations (FAO 2010, p. 46), Jatropha oil has different chemical properties than diesel fuel and burns at a different temperature. Consequently, all generators and engines have to be reconfigured to use Jatropha oil.

Crude Jatropha oil has several beneficial characteristics. Specifically, it stores well because it is fairly thick and low in free fatty acids. Unsaturated fatty acids with a high iodine value give it the capacity to remain liquefied at colder temperatures. It also ignites easily because sulphur content is low and, as such, emits lower levels of

Table 4.1 Comparison of the characteristics of fossil diesel to pure Jatropha oil

	Diesel oil	Oil of Jatropha curcas seeds
Density Kg/l (15/40 °C)	0.84–0.85	0.91–0.92
Cold solidifying point (°C)	−14	2.0
Flash point (°C)	80	110–240
Cetane number	47.8	51.0
Sulphur (%)	1.0–1.2	0.13

Source: Adapted from – Properties of Jatropha Oil. Food and Agriculture Organization of the United Nations – FAO (2010, p. 45)

sulphur dioxide (S02) when combusted as a fuel. However, on the negative side, the oil's high level of unsaturated oleic and linoleic acids makes it susceptible to oxidation when stored (FAO 2010, p. 45). Table 4.1 provides a comparison of the physical properties of Jatropha oil and diesel oil.

Floreana Pilot Project Results

Following the initial 2008–2011 development phase, the project expanded to provide Floreana with 100 percent renewable energy. Since February 2011, a hybrid solar and biofuel system has been providing 69 kW (kWel) of electrical power. This has successfully provided renewable energy for both inhabitants and tourists on Floreana Island (ENERGAL 2014). A second project has been established on Isabella Island, which has the second lowest resident population in the Galapagos.

Ecuador's Jatropha harvest increased fivefold from the initial 2009 and 2011 production stage to 2013 in order to meet biofuel demand on Floreana and the generators on Isabella Island. In 2012, a new Jatropha oil press was provided in Portoviejo to allow production increases. Jatropha harvest expanded from 137 tonnes in 2011 to 213 tonnes in 2012. Jatropha oil yields increased from 36,000 litres in 2011 to 50,000 L in 2012 (GIZ 2016). In addition, approximately 1200 more local families have benefited from income related to Jatropha production (GIZ 2016).

Future Opportunities for Biofuel Use

It is difficult to get a sense of the role of biomass in world energy mix. Based on energy sold through formalized markets, the US Energy Information Administration's (EIA) International Energy Statistics database suggests the average annual percentage of change in world energy consumption of biofuels from 2012 to 2020 and the use of renewable energy sources has doubled, as illustrated in Fig. 4.2.

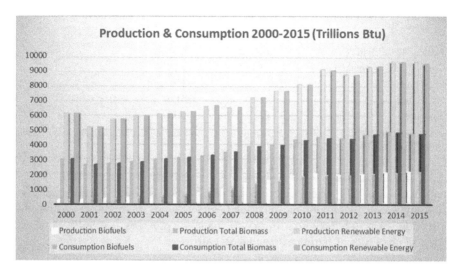

Fig. 4.2 Biofuels and biomass comparison to total renewable energy. Source: Adapted from US Energy Information Administration (2016b). Section 10: Renewable Energy. *Monthly Energy Review (MER)*, December 2016, pp. 149–165

Liquid biofuels represent 0.8% of the total energy consumption of Ecuador. This compares well with the Renewable Energy Market Analysis: Latin America - International Renewable Energy Agency (IRENA 2016) estimates that liquid biofuels represents just 0.2% of the total primary energy supply in total renewables. Although Jatropha production is relatively insignificant in the context of world energy production and consumption, biofuels and biomass energy sources contribute approximately 72% of residential and commercial consumption in the South American Andean States of Bolivia, Colombia, Ecuador, Peru, and Venezuela (IRENA 2016, p. 54). In fragile environments, such as the Galapagos, biofuel from Jatropha has the potential for greater use in a sustainable energy mix. The government of Ecuador's Zero Fossil-Fuel Initiative is ambitious, but progress is slowly being made. Both Floreana and Isabela Islands are operating generators running on Jatropha oil together with photovoltaics and wind energy generation. The importance of the Galapagos Islands World Heritage designation means that the care and protection of marine and terrestrial environments is critical to the future of the Islands and achieving sustainable energy mix solutions are a critical part of this protection.

Next Steps

According to Carvajal (2012), if electricity from diesel generators is replaced with solar, wind, and biofuels, the next step is the gradual conversion of diesel engines to biofuel engines. There is potential for Jatropha oil biodiesel to eliminate a large portion of the diesel oil used within the borders of Ecuador, and specifically the amount shipped across 1000 km of ocean to the Galapagos Islands. A critical question with respect to the increased future use of Jatropha oil in the Galapagos as an alternative to marine diesel is the risk of spills. Both fishing and tourist boat engines can also be converted to the use biofuels. The research literature specific to biodiesel fuel spills in fragile land and marine environments suggests there is generally less risk with respect to biofuel biodegradable properties. This is important given the marine and terrestrial habitats and species of the Galapagos Islands.

While the "Galapagos Islands Zero Fossil-Fuel Initiative" is an important step in supporting the development of a sustainable energy mix in the Galapagos, there is growing energy demand for land-based tourism and Ecuadorian migration. Sustainable energy mix solutions are critical for the future of the Islands. But, new policies and further research, together with social, economic, and environmental action are all needed to address the effects of increased growth pressures on the Islands' natural and social capital.

Acknowledgements Carlos Alberto Jácome, Regional Energy Specialist. International Development Bank.

References

Carvajal P (2012) Galapagos Islands zero fossil fuel initiative. Ministry of Electricity and Renewable Energy – Ecuador - MEER 2012. Retrieved from http://www.irena.org/DocumentDownloads/events/MaltaSeptember2012/Pablo_Carvajal.pdf

Deutsche Gesellschaft für Internationale Zusammenarbeit (GIZ) Gmbh (2016) Climate protection through the use of renewable energies on the Galapagos Islands, with special focus on power generation using jatropha oil (ENERGAL). Retrieved from https://www.giz.de/en/worldwide/12701.html

ENERGAL (2014) Renewable energy for Galápagos. Deutsche Gesellschaft für Internationale Zusammenarbeit (GIZ) Gmbh

Federal Ministry for the Environment, Nature Conservation, Building and Nuclear Safety, International Climate Initiative (2015) Renewable energy resources for the Galapagos Islands. September, 2015. https://www.international-climate-initiative.com/fileadmin/Dokumente/landingpages/sairec2015/ 151001_19_ Galapagos_Web.pdf

Federal Ministry for the Environment, Nature Conservation, Building and Nuclear Safety, International Climate Initiative (2016) Power generation using jatropha oil on the Galapagos Islands. January, 2016. https://www.international-climate-initiative.com/en/projects/projects/details/power-generation-using-jatropha-oil-on-the-galapagos-islands-354/?printview=printProjectAsPdf

Food and Agriculture Organization of the United Nations (FAO) (2010) Chapter 3: Jatropa Cultivation: Integrated Crop Management, vol 8, 29p, http://www.fao.org/docrep/012/i1219e/i1219e02.pdf

Gruber G (2014) Pure jatropha oil for power generation on Floreana Island/Galapagos: four years experience on engine operation and fuel quality. J Energy Power Eng 8:929–938. David Publishing, Allersberg, Germany

Instituto Interamericano de Cooperación para la Agricultura (IICA) (2013) Experiencias exitosas en Bioeconomía: producción de aceite de piñon para el plan piloto de generación eléctrica en Galápagos, pp. 18–21. IICA, Montevideo, Uruguay

Instituto Nacional de Estadística y Censos (INEC) – Gobierno Nacional de la Repùblica del Ecuador (2015) Principales Resultados Censo de Población y Vivienda – Galápagos 2015. http://www.ecuadorencifras.gob.ec/documentos/web-inec/Poblacion_y_Demografia/CPV_Galapagos_2015/Presentacion_CPVG15.pdf

International Renewable Energy Agency (IRENA) (2016) Renewable energy market analysis – LatinAmerica. http://www.irena.org/DocumentDownloads/Publications/IRENA_Market_Analysis_Latin_America_2016.pdf

Lougheed LW, Edgar GJ, Snell HL (eds) (2002) Biological impacts of the Jessica oil spill on the Galápagos environment: final report v.1.10 Charles Darwin Foundation, Puerto Ayora, Galápagos, Ecuador

Ministerio de Electricidad y Energía Renovable Sub-Secretaría de Energía Renovable y Eficiencia Energética (2016) Ficha Informativa de Proyecto: K015 MEER - Producción de Aceite de Piñón para plan piloto de generación eléctrica en Galápagos. http://190.152.52.4/LOTAIP/planificacion/SEGUIMIENTO%20GPR%20OCTUBRE2016/PRODUCCIONDEACEITEDEPINON.pdf

Ministry of Electricity and Renewable Energy (MEER) (2012) Galapagos Islands Zero Fossil Fuel Initiative. Ministry of Electricity and Renewable Energy – Ecuador – Conference Presentation of Ministry Pablo Carvajal at Renewables and Islands Global Summit organized by International Renewable Energy Agency - IRENA. Malta, September, 2012

Ministry of Electricity and Renewable Energy (MEER) (2013) SE4ALL Rapid assessment and gap analysis of the energy sector RG-T1881-Ecuador, Sustainable Energy for all Americas - United Nations Development Program (UNDP), 127p

Quiroga D (2014) Ecotourism in the galapagos: management of a dynamic emergent system. George Wright Forum, George Wright Society, Hancock MI (Publisher) 31(3):280–189

Roundtable for Sustainable Biomaterials (RSB) (2012) A guide to the RSB standard. Independent and global multi-stakeholder coalition which works to promote the sustainability of biomaterials with NGOs and UN agencies. January, 10th 2017. http://rsb.org/pdfs/documents_and_resources/RSB%20Standards%20Guide.pdf

Strauss P, Braun M, Estrella R, Karres S, Klaus W, Manzano L, Moreno A, Rodriguez I, Samaniego A (2011) ISES 2011—Proceedings International Solar Energy Society. High penetration PV- battery-straight vegetal oil hybrid system for zero emission electricity generation on the Galapagos Islands, August 28—September 2, Kassel, Germany

US Energy Information Administration (2016a) International energy outlook 2016 with projections to 2040. May–June 2016, 290 p. Retrieved from http://www.eia.gov/outlooks/ieo/pdf/0484(2016).pdf

US Energy Information Administration (2016b) Section 10: renewable energy. Monthly energy review (MER), December 2016, pp. 149–165. Retrieved from http://www.eia.gov/totalenergy/data/monthly/pdf/mer.pdf

Chapter 5
Policies and Laws and Island Environments

Allan Ingelson and Christopher Phillip

Introduction

During the last decade, the generation of electricity from renewable energy (RE) resources has increased significantly in countries around the world. Government policies and laws can play an important role in facilitating the development of RE resources. As an extension of discussions at the 2014 World Energy Summit in the Galapagos Islands about the obstacles to renewable energy on remote islands, in this chapter, we consider examples of RE technologies that have been successfully deployed on islands in different regions of the world to generate electricity. Then we will examine the obstacles to using RE to produce electricity and the government policies and laws that have been used to promote RE development. In some countries governments have employed a combination of different programs and incentives to facilitate electricity production from RE.

Geothermal and wind provide examples of two RE resources that have a proven track record of electricity production on islands. Recently solar energy is emerging as increasingly important RE source for island environments. As has been the case with other types of energy, the potential for developing RE on islands depends upon the availability and cost of competing sources of energy such as fossil fuels and the availability and intensity of the specific RE resource itself such as wind. Historically in countries without domestic fossil fuel reserves, where RE resources are available, governments such as Japan have created policies and adopted laws to promote RE

A. Ingelson (✉)
Faculty of Law, University of Calgary, Canadian Institute of Resources Law, Calgary, AB, Canada
e-mail: allan.ingelson@ucalgary.ca

C. Phillip
Student-at-law, British Columbia (B.C.) Hydro, Vancouver, BC, Canada

development and provide national energy security. Today in response to the anticipated effects of climate change, the United Nations and more governments are assisting with facilitating RE development. With regard to much smaller islands, in light of climate change, the United Nations Education, Scientific, and Cultural Organization (UNESCO) International Hydrological Programme initiated a groundwater resources assessment program in 2004, to increase our understanding of the effects of climate change on global groundwater resources and promote more sustainable groundwater management to protect groundwater.[1] UNESCO has identified a group of 52 developing countries on islands in the Pacific Ocean, Indian Ocean, South China Sea, Caribbean, and Africa that collectively have a population of 63.2 million residents and a gross domestic product of US $575.3 billion.[2] Economic challenges faced by governments of the Small Island Developing States include high energy, infrastructure, and transportation costs that are an important consideration when evaluating the economic viability of developing RE projects.[3] The availability of financing for RE projects either from national governments or through international assistance is an important factor in the level of RE development.

An important consideration is the variability in the generating capacity of some RE technologies such as solar and wind development and the availability of alternate energy sources that can be brought online to maintain a reliable electricity supply for customers at night or when there is a reduction in wind velocity. Unlike RE resources with a variable energy output, geothermal energy has the advantage of providing consistent baseload electricity generating capacity which in part explains the more extensive development of geothermal energy on some islands where the geothermal resource is available than solar or wind.

The size of the island and the associated population where there is RE production vary significantly, and this can influence the electricity demand and the size of the energy market required to meet consumer needs. At one end of the scale, there are small islands where a few 100 people reside and at the other end of the scale island continents such as Japan, the United Kingdom (UK), New Zealand, and Australia host millions of residents that create much larger markets for electricity. The UK provides an example of an industrialized nation comprised of large islands with a substantial population that has made offshore renewable wind development a central focus of its policy and laws to reduce its dependence on coal and the associated carbon emissions and achieve its ambitious climate change targets.[4]

---

[1] UNESCO/Division of Water Sciences, "GRAPHIC Groundwater and Climate Change, Small Island Developing States (SIDS)," SC-2015/WC/29, at 1, ihp@unesco.org – www.unesco.org/water/ihp.

[2] United Nations. International year of Small Island Developing States – 2014, at http://www.un.org/en/events/islands2014/index.shtml#&panel1-1>.

[3] Supra, note 2, p. 3.

[4] Glen Plant, Offshore Renewable Energy Development in the British Islands – Part 1, 2 RELP (2013) 120–142.

Solar Energy: South Pacific Islands

Solar energy appears to be a more environmentally benign RE technology than geothermal and wind. Notwithstanding the substantial cost of diesel generation, the majority of small island nations in the world still rely heavily on fossil fuels. However, in light of technological innovation in solar energy in recent years and a substantial reduction in the cost of solar photovoltaic (PV) systems in the last decade, during the last couple of years, there have been several small islands in the South Pacific on which solar systems have been installed with international financial assistance in the form of loans and grants. Prior to 2012, the residents in Tokelau Island in the South Pacific relied on three diesel power stations for their electricity needs.[5] With financial assistance from the government of New Zealand, today, three solar PV systems supply 100% of the electricity demand on the island atolls, replacing electricity generated by diesel power plants. Kiribati in the South Pacific provides another recent example of solar energy development on small remote islands where 2000 residential units and 100 community solar systems have been installed in villages on 18 islands as a result of funding from Japan and the European Union.[6] In addition to securing financing for solar RE projects on remote islands, one additional challenge on islands subject to hurricanes is the damage to rooftops and solar PV panels from strong winds. The potential for hurricane damage should be an important planning consideration for the placement of panels to minimize damage in hurricane prone areas.[7] As water is a critical resource on numerous remote islands and limited nonsaline water resources on numerous remote islands, the energy water nexus should be an important consideration when planning for solar energy projects on remote islands. As water is required to clean the solar panels for optimal electricity generation, water usage is a consideration for solar projects on islands with limited water supplies. In addition to electricity generation, solar energy has been used for many years in solar hot water systems, a second benefit from developing this RE resource.

Residents of the isolated Hawaiian Islands face significantly higher residential electricity rates than states on the US mainland. One issue that has arisen in the isolated island state of Hawaii is the opposition from the state electricity utility to increasing the number of rooftop solar installations in light of the potential decrease in the number of utility ratepayers contributing to the costs of maintaining the electricity grid. In 2015, the regulator called the Hawaii Public Utilities Commission (PUC) filed a ruling to make it more attractive for island homeowners to install rooftop solar systems to generate electricity.[8] The PUC has recommended a "self-supply tariff" (SST) and a "grid-supply tariff" (GST) to future rooftop solar electricity installations. The SST will facilitate an expedited review of rooftop solar

---

[5] Rocky Mountain Institute, www.cleantechnica.com.../an-island-tokelau-powered100.
[6] www.worldviewofglobalwarming.org/pages/solarinkiribati.ph.
[7] www.caymannewresident.com/solar.
[8] Travis Holum, "Hawaii May Have Just Given Energy Storage a Huge Boost," October 22, 2015, at http://fool.com/investing/general/2015/10/22/hawaii-may-have-just-given-energy-stor.

installations, and the GST will allow those residents generating their own electricity from rooftop solar to avoid paying higher retail electricity rates paid by those people without rooftop solar. In addition all homeowners will pay $25 US per month as compensation to the utility for providing grid infrastructure.[9] Hawaii provides one recent example of how changes in an energy regulatory system can promote increased development of rooftop solar production.

Geothermal Energy: Iceland, Indonesia

Geothermal energy is one of the RE resources that has been developed on numerous islands for several decades. Favorable island geology which includes volcanic rocks is an important factor in determining whether economic development of the geothermal resource is viable. More than four decades ago, the lack of domestic fossil fuels on the remote island of Iceland combined with the favorable geology prompted the government to develop geothermal energy for electricity production and space heating. By 2015, approximately 665 MW of geothermal electricity generation capacity was installed in Iceland, representing 30% of the country's electricity generation (Bertani 2015). The Philippines is now the second-largest geothermal electricity producer in the world with an installed capacity of 1.96 GW.[10] In the Hawaiian Islands, the installed geothermal capacity is 38 MWe.[11] Geothermal supplies 25% of the electricity consumed on the Big Island of Hawaii.[12] Other remote islands that have developed geothermal energy include the Azores off the Portuguese mainland where geothermal energy supplies 42% of the electricity consumed on Terceira Island and more than 22% of the total electricity demand on the remaining islands in the archipelago.[13] Fifteen percent of the electricity generated in Costa Rica is from geothermal energy.[14] On the island of Guadeloupe, France, 16 MWe of installed capacity from geothermal has been developed that provides up to 20% of the total electricity supply for the island.[15] In Russia, along the Kamchatka Peninsula and on the Kuril Islands, there is 82 MWe of installed capacity from the geothermal resource.[16] Geothermal energy in New Zealand supplies 13% of the electricity needs of the island nation.[17]

In addition to electricity generation, geothermal energy in Iceland provides 90% of household space heating requirements (Orkustofnun n.d.-a). The Icelandic geothermal

---

[9] Ibid.

[10] www.geothermal-energy.org/electricity-generation/philippines.html.

[11] www.geothermal-energy.org/electricity-generation/usa.html.

[12] Ibid.

[13] www.geothermal-energy.org/electricity-generation/portugal_the_azores.html.

[14] www.geothermal-energy.org/electricity-generation/costa-rica.html.

[15] www.geothermal-energy.org/electricity-generation/france-guadeloupe.html.

[16] www.geothermal-energy.org/electricity-generation/russia-kamchatka.html.

[17] www.nzgeothermal.org.nz/elec_geo.html.

experience has also revealed that there are multiple other beneficial uses of geothermal resources that have increased the economic benefits associated with geothermal energy development. Substantial revenue has been collected from tourists visiting the Blue Lagoon hot springs spa, aquaculture, and cosmetics activities.

The Icelandic geothermal development experience has revealed that the subsurface geothermal water needs to be reinjected to provide for sustainable energy production. If the energy withdrawal rate exceeds, the rate at which the energy from the Earth's core reaches the subsurface reservoir through conduction, then geothermal resource output (and energy efficiency) will gradually decline. Alternatively, when steam or water is directly withdrawn from a geothermal energy reservoir and not reinjected or replenished, output will decline over time.

To attract investment in geothermal energy development, certainty in the legal framework surrounding geothermal rights is required to provide prospective geothermal developers with sufficient incentive to make the significant capital investment in a geothermal power plant. In Iceland one key factor that has contributed to the success of geothermal development on the remote island is a supportive government policy and an associated legal framework that assigns geothermal energy ownership rights to surface land owners in the same manner that subsurface mineral rights are allocated to minimize disputes that could delay and frustrate geothermal development. The *Act on Survey and Utilization of Ground Resources* (Law No. 57/1998) covers any Icelandic "resource" found in the ground, at the bottom of rivers and lakes, or at the bottom of the sea within netting limits. A "resource" is defined in the act as "any element, compound and energy that can be extracted from the earth, whether in solid, liquid, or gaseous form, regardless of the temperature at which they may be found." While the ownership of geothermal resources is based on the ownership of land, the utilization of geothermal resources is subject to a license from the energy regulator Orkustofnun, a government agency whose primary responsibility is advising the Icelandic government on energy matters (Orkustofnun n.d.-b). Therefore, while a landowner may also have ownership over a geothermal resource beneath their land, that person does not have the right to develop geothermal energy for electricity production without securing an utilization license from the Orkustofnun. One important advantage of Iceland's regulatory framework is that it allows landowners to benefit from commercial development of the geothermal resource discovered beneath their land, although ultimate jurisdiction over geothermal development is granted to the Orkustofnun such that utilization licenses can be awarded in a manner that does not deplete the underlying geothermal resource. Iceland provides an example of how the adoption of a legal framework that provides resource development certainty to developers and benefits to proximal stakeholders has facilitated this type of RE resource rather than providing subsidies. The stimulus for RE development is providing a clear legal definition of geothermal rights and designating ownership rights in situations where there is the potential for resource ownership dispute.

Indonesia consists of an archipelago of more than 17,500 islands situated along the Pacific Rim of the volcanically active "Ring of Fire," over 40,000 km of active subsea volcanoes and ocean trenches that wraps around the major land masses on either side of the Pacific Ocean. Based on its geology, Indonesia has significant potential for increased geothermal development as the country is estimated to host more than 40% of the world's potential geothermal resources. As with the Government of Iceland, the central Indonesian government is keen to promote development of its geothermal resources in light of the rapidly increasing electricity demand from both consumer and industry groups. In 2003, Indonesia adopted its first *Geothermal Law* (Law No. 27/2003), which formalized the regulatory regime for what had been a patchwork of geothermal development laws in the country. A periodic call for tenders for surveyed geothermal fields was issued by the central government to prospective geothermal developers (IRENA 2015i). The RE developers were then to compete in a closed-bid auction for geothermal business permits (IUPs) that would grant the right to exploit the subsurface geothermal resource in a given area. Developers successful in securing an IUP would still be responsible for negotiating surface access for development. In 2004, the Ministry of Energy and Mineral Resources (MEMR) issued its "Blueprint for Geothermal Development in Indonesia" that includes a stated goal of 6000 MW of geothermal capacity by 2020. In 2005, the Directorate of Geothermal Energy (formerly the Directorate of Geothermal Enterprise Supervision and Groundwater Management) was created and made responsible for overseeing research on Indonesia's geothermal energy resource potential and future development. In 2012, the MEMR initiated a feed-in tariff (FIT) program[18] for geothermal energy to provide economic incentives for additional geothermal electricity development. Operational geothermal plant facilities were forced to sell their electricity to local utilities at the spot price market price. More than 1335 MW of geothermal generation capacity has been installed in Indonesia by the end of 2012 (Meier et al. 2015).

Despite the FIT program was aimed at promoting geothermal energy development, and $200 million in guaranteed loans was provided by the Ministry of Finance to assist geothermal electricity developers due to the substantial upfront capital costs to construct a geothermal generating plant, by 2013 geothermal development in Indonesia had largely stalled (Meier et al. 2015). In an attempt to revive geothermal electricity development, the Indonesian government announced amendments to its geothermal legislation in a new *Geothermal Law* (Law No. 21/2014), that replaced the old *Geothermal Law* (Law No. 27/2003) (IRENA 2015j). The new law eases the regulatory burden on prospective geothermal developers in number of ways (Sastrawijaya et al. 2014). Under the 2003 *Geothermal Law*, no distinction was drawn between direct (heating purposes) and indirect (electricity generation) utilization of geothermal energy – an IUP could be issued for either type of development by either one of the MEMR, regional governor, or a local mayor depending on where the geothermal working area was located. Under the 2014 *Geothermal Law*, the IUP scheme was divided into direct utilization licenses for direct geothermal

---

[18] An ongoing offer to energy developers that provides a connection to the transmission grid, costs borne by utilities, and a premium price for the electricity generated from the geothermal project; there will be more discussion of FIT programs later in the chapter.

development and geothermal licenses for indirect geothermal developments. Regional governors and local mayors would still be allowed to issue direct utilization licenses; however, geothermal licenses for electricity generating stations can now only be obtained from the MEMR. This was done in an effort to centralize the tender process, as well as (presumably) to reduce the opportunity for regional government officials to demand a bribe in a country with a reported corruption problem – currently ranked 107 out of 174 countries in Transparency International's 2014 Index (Transparency International 2015). Under the new *Geothermal Law*, regional governments are now entitled to a set a "production bonus" as recompense for geothermal facilities in their jurisdiction, which again (presumably) limits bribes or kickbacks local officials may request from geothermal energy developers.

One of the more controversial elements of the most recent *Geothermal Law* adopted in Indonesia is an exception that provides geothermal activities will no longer be considered to be "mining activities," which under Indonesian forestry law are prohibited in conservation forest areas (Sastrawijaya et al. 2014). This effectively opens up extensive tracts of Indonesia for new geothermal development at the risk of damaging the few pristine forest regions Indonesia that remain in. The country provides an example of a jurisdiction in which seemingly common goals of the environmental movement are at odds – environmental protection of pristine forest regions versus geothermal energy development with the potential to offset significant amounts of GHG emissions from a predominantly coal-powered grid. It will be interesting to see if a balance between these two goals can be struck as geothermal energy development continues in coming years. Another environmental challenge that has arisen from geothermal electricity development in the island nation of the Philippines is the potential for water contamination. Depending upon the geology of the specific area and chemistry of the rocks where geothermal wells are drilled, steam and fluids that are released may contain metals such as arsenic. At some sites this is an important environmental and health and safety risk that needs to be carefully investigated and managed if geothermal energy development proceeds.

Wind: Samso Island, Denmark and Ramea, and Prince Edward and Haida Gwaii Islands, Canada

Samso Island provides one example of successful wind and biomass energy development. The island that is located 15 km off the Jutland Peninsula has a total area of 114 km^2 and approximately 3800 full-time inhabitants (Visit Samso 2015). Historically the residents encountered an unreliable electricity supply and higher energy costs than on the mainland, problems that are similar to those encountered on other remote islands. Historically there was the lack of a transmission connection to a central power grid and dependence on imported fossil fuels such as diesel for both transportation and electricity generation. In an attempt to solve these problems, Samso, the energy developer, bid in a Danish government contest to become "Denmark's Renewable Energy Island" in 1997 (Saastamoinen 2009a, b). In winning the contest, Samso gained access to a variety of government subsidies and began collaboration with a host of renewable energy industry players and energy consultants to redevelop the island's RE systems. The initiative focused on developing a 100% renewable energy-powered pilot island within 10 years that could demonstrate energy self-sufficiency and serves as a model for other island environments elsewhere. The plan focused on reducing energy consumption and increasing energy efficiency, expansion of a district heating supply system, and both on and offshore wind turbines to generate electricity. In 2005, Samso achieved its energy goal 2 years ahead of its original timeline.

Samso Island today has eleven 1 MW onshore wind turbines, ten 2.3 MW offshore wind turbines, three straw-powered heating plants, and one solar and one wood chip heating plant (visit Samso 2015). The island generates more electricity than it uses and sends surplus electricity to the main Danish grid for sale in the Nord Spot Pool via a new underwater transmission cable that connects to the mainland 15 km away. As part of gaining local stakeholder support for the redevelopment of energy on the island, landowners were offered an opportunity to buy a stake in the wind turbines being erected and collect a share of the revenues from the sale of electricity to the main power grid. A district heating system fueled by hay grown on the island, pipes hot water to most of the island residences. Although fossil fuel-based transportation fuels have proven more difficult to replace entirely, the island actively promotes electric vehicles among its residents, and the government has plans to use landfill gas from its local waste management facility as energy for the ferries that run between the island and the Danish mainland. The entire project is estimated to cost about $80 million USD – approximately 20% of which came in the form of government subsidies (Cardwell 2015).

Issues have arisen in recent years with respect to the maintenance costs of the wind turbines and lower than expected revenues from electricity sales based on Nord Spot Pool oversupply; however, overall the redevelopment of Samso Island is widely regarded as a success story. While certain features of Samso Island, for example, its supply of biomass from local agriculture, may make the Samso model not entirely replicable in other island environments, the Danish government and Samso's local government actively promote the island today as a tourist destination for those interested in renewable energy and research base for renewable energy developers. The Samso Energy Academy was opened in 2007 and now houses a collection of renewable energy researchers, public education, and training programs (Saastamoinen 2009a, b).

Ramea Island and Prince Edward Island in Canada provide two additional examples of successful wind energy development. Ramea is located in the Atlantic Ocean of the southern coast of Newfoundland in Eastern Canada. The island has an area of a little more than 3 km^2 and a population of approximately 600 year-round inhabitants. In 2004, the island was selected by the Canadian federal government department Natural Resources Canada (NRCan) as the test site for a proposed trial wind-hydrogen-diesel hybrid power project based on its wind energy potential, local community approval of wind development, and rising costs associated with importing diesel fuel for electricity generation on the island (NRCan 2014).

The project has been jointly operated by NRCan and Newfoundland and Labrador Hydro (now a subsidiary of the provincial Crown corporation Nalcor Energy) with approximately CAD $12 million (USD $9.3 million) in funding from both the federal and provincial governments.

The goal of the Ramea Island project is to satisfy the total electricity requirements of a small number of residents during low-demand periods through a combination of direct wind power generation and combustion of stored hydrogen gas (NRCan 2014). The production of hydrogen gas is an energy-intensive process that is often energetically less favorable than other technologies in that it requires more energy to produce a molecule of hydrogen than is released when the same molecule is combusted. The Ramea Island project has been designed to electrolyze water to produce hydrogen using energy from the wind power supply during periods when the wind power supply exceeds the net power demand on the island. In this way, electricity generated by combustion of hydrogen gas during off-peak wind supply periods could then be used to offset the island's diesel generation requirements. The trial project targeted a wind-hydrogen power supply to satisfy the island's low-demand period electricity requirements, with diesel generators still being used to supply peak electricity demands during the cold winter months. The island does not have any nearby transmission capacity to the provincial mainland in Labrador that excess electricity from its wind turbines could be transported and marketed to. As a result, electricity storage in the form of hydrogen gas presents one appealing option for remote islands such as Ramea if proven to be technologically feasible. While the system was originally completed in 2012, Nalcor Energy has been operating the project throughout the demonstration phase and studies remain ongoing (Nalcor n.d.; Islam 2012). In January 2014, the federal government announced an additional CAD $2.3 million (USD $1.8 million) in federal funding for ongoing support of the wind-hydrogen energy project (ACOA 2014). The long-term goal of the Ramea Island wind-hydrogen-diesel hybrid power project is to create an economicly viable wind energy storage system for remote, diesel-powered communities on islands.

Another successful wind energy project on the east coast of Canada funded by Canada's National Government has been developed on Prince Edward Island (PEI). The island has experienced significant agricultural development, and the commercial fishing and tourism industries make an important contribution to the island economy. PEI is a federal government wind energy research test center. Tourism "has increased exponentially" in areas of PEI where wind farms have been constructed. For example, the number of visitors at one of the wind farms on the North Cape "exceeds 60,000 people" annually since the opening of the wind farm and seasonal revenues from the gift shop and restaurant generating $260,000 annually from May through October.[19]

A third large wind energy project proposed on the Haida Gwaii (HG) Islands, on the west coast of Canada in northwestern British Columbia formerly known as the Queen Charlotte Islands, illustrates some of the challenges faced by wind energy

---

[19] http://www.canwea.ca/images/uploads/Fle/New_releases/CanWEA_Release_-_October25(1).pdf; http://www.canwea.ca/images/uploads/File/North_Cape2.pdf.

developers in coastal areas. The HG islands include two major islands – Graham Island in the north and Moresby Island in the south – along with a collection of smaller islets for a collective landmass of just over 10,000 km^2. As in the Galapagos Islands in Ecuador, the remote HG islands encompass a national park and are both ecotourism destinations. The HG islands are the home of indigenous peoples known as the Haida Nation. Approximately 4500 people live on the island today, with 45% of the population consisting of members of a variety of indigenous peoples. The UNESCO World Heritage Site SGang Gwaay is located in Gwaii Haanas National Park Reserve. The islands are considered to be "a world-class destination for the adventurous traveler…which feature unique temperate rainforests, pristine uncrowded beaches, abundant wildlife and renowned ancient Haida village sites."[20] The HG Islands are not connected to BC Hydro's mainland integrated grid, and as such are serviced at present by BC Hydro's Non-Integrated Area services, which manage the generation and distribution of electricity in many of British Columbia's remote northern communities. Graham Island's North Grid is currently serviced by a BC Hydro-managed diesel-powered generating station, while Moresby Island's South Grid is powered by a privately owned hydroelectric generating station that is backed up by another BC Hydro-managed diesel generating station.

In BC, there is a large, steady supply of baseload electricity from hydroelectricity that accounts for over 95% of electricity generated in the province. In June 2008, BC Hydro a utility owned by the provincial government issued a Clean Power Call Request for Proposals (RFP), which after a round of amendments culminated in a call for upward of 3000 GWh of annual electricity from designated renewable energy technologies (BC Hydro 2015). By August 2008, 168 projects of interest had been registered by BC Hydro, although by the proposed submission deadline of November 2008, only 68 proposals were received. These 68 proposals, however, still represented a collective 17,000 GWh annual electricity output. After a comprehensive evaluation process that took until August 2010 to complete, BC Hydro selected 27 projects from throughout the province and offered electricity purchase agreements (EPAs) (BC Hydro's PPA equivalent) to successful project proponents. These projects included 19 run-of-the river projects, 6 wind projects, 1 storage hydro project, and 1 waste heat project (BC Hydro 2015).

A large wind energy project called the NaiKun Wind Energy Project (NaiKun) was proposed by a private corporate wind energy developer to construct 110 turbines in a 396 MW offshore wind farm that would be located in the windy channels of the Hecate Strait that runs between the HG islands and mainland BC. NaiKun's CAD $2.5 billion wind project would connect to an onshore substation on Graham Island and replace all of the diesel-generated electricity on the North Grid. All surplus electricity would be transported to BC Hydro's mainland grid via a 100 km subsea transmission line (NaiKun 2015). Realizing the unique stakeholder of concerns of coastal British Columbia and the recognized territories of the indigenous First Nations bands on Haida Gwaii, the wind developer completed an extensive consultation process with indigenous residents that ultimately culminated in

---

[20] www.haidagwaitttourism.ca.

the signing of a tentative agreement with the Council of Haida Nation (CHN) that would provide for the CHN to assume a 20–40% stake in the project dependent on financing (Simms 2011) and the environmental assessment (EA) of the of the island wind energy project.

Despite the completion of the provincial environmental impact assessment (EA) process for the wind energy project by the energy developer and the issuance of BC EA certificate by the provincial government in 2009, PPA tenders for successful projects under the Clean Energy Call RFP were awarded to smaller projects (less than 100 MW) in capacity, signifying support from BC Hydro for smaller, more distributed projects rather than large, single proponent projects. In response to the proposed major wind energy project, BC Hydro, a Crown Corporation in March 2010, informed the NaiKun wind energy developer that its proposal had been eliminated from consideration for a power purchase agreement (PPA) due in part to an excessive risk profile. Despite the tentative agreement NaiKun had reached with the CHN, a number of smaller groups with the Haida Nation voiced opposition to the NaiKun's project despite council's consent to the project, citing concerns with potential risk to the local subsistence fishery in the Hecate Strait and the CHN's ability to make interest payments on the hefty loans that would be required to buy-in to the project as shareholders (Simms 2011).

Tidal Energy

Unlike the other RE resources, solar, geothermal, and wind, that we have discussed, the development of tidal energy has been much more restricted. There has been limited electricity generation from facilities in Normandy, France, Scotland, and in the province of Nova Scotia, Canada. Nova Scotia Power built the first and only tidal generating station in North America that can produce up to 20 MW.[21] Tidal energy development is currently an active area for experimentation and technological innovation. As a result the policies and laws that have been adopted by governments to facilitate solar, geothermal, and wind energy development are at a more advanced stage than those that may be developed to facilitate tidal energy production.

Laws and Policies to Support Renewable Energy

We have now considered a few examples of successful RE projects on remote islands and some of the obstacles to development. In the next section of the chapter, we will examine the basic policy and legal approaches that have been employed by governments to promote national RE development in a much broader context than

---

[21] Fundy Ocean Research Center for Energy (FORCE), www,fundyforce.ca.

remote islands. Should a government decide to focus on promoting RE development on remote islands, then the following fundamental policy and legal approaches can form the basis for stimulating RE electricity production in a specific area, when the policies are tailored to the challenges unique to remote island environments such as smaller electricity markets than those in cities on the mainland and higher electricity costs due to higher diesel generation costs.

While minor differences exist in jurisdictions with respect to policies designed to promote RE development and the laws that provide the framework to implement the policies, the following three general approaches have been adopted by numerous governments to facilitate RE development at the national level:

1. Feed-in tariff (FIT) programs;
2. Renewable portfolio standard (RPS) systems;
3. Periodic calls for tenders from renewable energy developers.

In some nations a combination of two or three above approaches have been incorporated into legal systems designed to promote RE. In light of the higher energy costs on remote islands due in part to higher transportation costs for fuels such as diesel and LNG and smaller populations and markets, these factors should be taken into consideration by policy and law makers when creating systems that might increase RE development on smaller remote islands. We will now examine the three basic vehicles used by governments to promote increased RE development at the national and regional levels.

Feed-in Tariff (FIT) Programs

Germany, Denmark, Japan, the USA, Canada, and Ecuador provide a few examples of countries where governments have adopted FIT programs. Under this type of program, the government typically provides an ongoing open offer to RE developers that their electricity projects will be connected to the main transmission grid once operational with the cost usually borne by the utilities, not the RE developers. Frequently there is a right of first purchase (ROFP) or priority grid access to ensure that all generated electricity is sold. Often, premiums (i.e., tariffs on the purchasing utility or downstream customer) are offered for this electricity which guarantees a minimum price per kilowatt-hour (kWh) above the average grid electricity price that varies by RE technology. The FIT rates are often set to correspond to calculated break-even costs for certain RE technologies to assist project developers in recovering upfront capital costs that are often substantially higher than those for electricity producers with plants that operate on fossil fuel-based sources. These premium rates are often offered to individual project developers for a period of 15–20 years that are guaranteed under either a contractually binding power purchase agreement (PPA) or, in certain common law jurisdictions, the administrative law principle of legitimate expectations. After this period is completed, all electricity from these projects is subject to normal market electricity prices in the particular jurisdiction or in a combined power pool. A key feature of numerous FIT programs is the periodic rate amendments made as certain RE technologies become less costly or become a specific focus area for a government.[22]

Germany

During the last 25 years, Germany has been a pioneer in the development of FIT programs and has continually revised its RE policy and legislation as technological innovation and experiences from the deployment of RE wind and solar technology provide insight to other governments and regulators interested in promoting RE development. Germany has been very successful in promoting coastal wind energy development and continues to be a world leader in developing and implementing RE technologies.

The 1991 German *Electricity Feed-In Law* (*Stromeinspeisungsgesetz*) required grid operators to connect RE projects to the central German electricity grid (IRENA 2015e). In addition, the public utilities that operated the grid were required to pay premium prices via a FIT for the electricity supplied from these RE power plants. All costs of the FIT incentive were recovered by the public utilities that operated the German grid at the time from downstream customers via a renewable energy surcharge. The premiums offered to RE developers fluctuated annually based on the average per kWh cost of all electricity sold via the public electricity grid in the previous year. Wind power plants and solar PV plants received the highest remuneration with a 90% premium (i.e., 1.9 times the average per kWh cost of all electricity sold the previous year), followed by hydropower, biomass, and biogas power plants smaller than 500 kW at 75%, rising later to 80% in 1998 (IRENA 2015e). Hydropower, biomass, and biogas power plants larger than 500 kW but smaller than 5 MW received a 65% premium, although projects larger than 5 MW were not eligible for any premiums (IRENA 2015e).

While no formal contracts in the form of PPAs were ever signed with RE developers that would guarantee the anticipated premiums for their electricity, the German constitutional principle of legitimate expectations (*Vertrauensschutz*) provided some legal certainty that the premiums would not be eliminated altogether or unfairly reduced (Thomas 2000). Similar to the English common law doctrine of legitimate expectations, the principle essentially provides that any representations made by a public authority may be reasonably relied on and allow for an action against the public authority should this representation be relied upon to their detriment.[23] In 1998, the *Electricity Feed-in Law* was amended as heavy develop-

---

[22] For example, see offshore wind energy development in Germany or geothermal energy development in Indonesia.

[23] The doctrine of legitimate expectations, a common principle of administrative law in many jurisdictions, has evolved differently via case law in common law jurisdictions such as Canada, the USA, England, and Australia to either broaden or narrow the doctrine along with the procedural

ment of wind farms around the coastal regions of Germany had placed an unfair financial burden on utilities to not only connect these new developments but pay a substantial premium on the electricity generated from these projects. Therefore, a cumulative 10% cap was introduced limiting the amount of electricity from RE sources that had to be paid according to the law. The total burden of the law was limited to individual utilities and their customers.

The *Electricity Feed-in Law* is widely considered to be the driving force behind the rapid expansion of RE development in Germany, particularly fast-paced wind farm development along parts of the German coastline. While the 1998 cap attempted to control the rising cost of electricity generation in certain areas in the country and encourage other types of RE development (i.e., non-wind), the rapidly decreasing cost of wind power made further wind project installations far more desirable than other RE technologies at the time. In 2000, with much of northern Germany already at or quickly approaching the 10% cap based on existing wind installations, the German government revamped its renewable energy regulatory framework based on the same general principles as the *Electricity Feed-In Law*, but it incorporated lessons learned over the previous decade of experimentation with RE development.

In 2000, the *Renewable Energy Sources Act* (*Erneuerbare-Energien-Gesetz* (EEG)) replaced the *Electricity Feed-In Law* of 1991 with the goal of doubling the share of electricity produced from RE by 2010 (IRENA 2015f). The EEG shifted the legal obligation to provide grid access for new RE installations from the utilities to grid operators. Under the legislation adopted in 2000, developers of new RE projects had to pay the cost of physically connecting their projects to the main transmission grid, although tertiary costs related to the reinforcement of transmission infrastructure to handle increased loads in certain areas of the country were to be incurred by the grid operators. The EEG also attempted to address the problem of an inequitable distribution of financial burdens among regional utilities that emerged under the *Electricity Feed-In Law*. Rather than imposing a maximum cap on the amount of electricity, RE regional utilities were required to buy; an ex-post quota system was implemented that required all electricity from renewable energy projects to be bought and incorporated into the grid-like generation from any other sources, but every 3-month period or quarter, the net RE purchased under the EEG by all utilities in Germany would be totaled and then added to the cost of EEG electricity shared among all utilities equally or proportionately – whether or not they generated or transmitted any electrons from renewable energy sources during that period or not. The ex-post quota system, in which utilities only learned what their total EEG bill for the quarter would be after the quarter ended, effectively allowed for the best RE sources in Germany (i.e., wind in northern coastal Germany) to be developed without financially crippling utilities in certain regions. While the actual costs incurred by some utilities and grid operators associated with load management and increased physical infrastructure requirements for EEG electricity were not explicitly known at that time, the primary effect of the ex-post quota system was to

---

and substantive rights offered under it.

redistribute the growing costs of purchasing electricity from RE sources equally among German utilities.

Tariffs for different RE technologies under the EEG were also adjusted from those offered under the *Electricity Feed-In Law* to reflect actual generation costs associated with each technology rather than on a premium of the average cost of grid-generated electricity. For example, new onshore wind project remuneration was fixed at € 8.4¢/kWh (USD 10¢/kWh) rather than 90% of the previous year's average grid per kWh cost. Remuneration for electricity from individual plants would be fixed for 20 years after generation begins for every technology except wind, which would have a fixed total kWh production cap at a higher premium, after which electricity from the individual installations would receive a reduced remuneration for the remainder of the initial 20-year period. The fixing of remuneration rates (outside of those for wind developers) was done with the intention to encourage RE generators to reduce their operational costs, as increasing currency inflation rates meant that the actual value of remuneration paid decreased over time in real terms. Additionally, from 2002 onward, the remuneration to newly commissioned plants has been decreased annually in order to further encourage cost reductions – by 5% annually for new solar PV installations, 2% for wind power plants, and 1% for biomass-fueled plants.

The German parliament reevaluates the EEG every 2 years on the basis of a report that is prepared by the Ministries of Economics and Technology (MET) in order to assess further efficiency gains and remuneration adjustments based on RE development to date.

After such a biannual review resulted on January 1, 2009, the EEG was amended. Increased costs for wind turbine raw materials such as copper and steel led the German parliament to offer increased tariffs for both onshore and offshore wind energy development after lobbying from the domestic wind energy industry. The tariff for new onshore wind projects had decreased to € 8.03¢/kWh (USD 9¢/kWh) in line with the 2% annual remuneration reduction offered to new projects, although this tariff offering had become prohibitively low to entice new wind farm development given the increasing cost of materials required for the manufacturing of turbines. As a result, the 2009 *EEG* amendment increased the tariff for onshore wind to € 9.2¢/kWh (USD 11¢/kWh) for the first 5 years of operation, decreasing to € 5.02¢/kWh (USD 6¢/kWh) after that (IRENA 2015g). Also, realizing that the annual remuneration decrease for new onshore wind energy developments may have been too large for developers to meet the targeted operational cost reductions, annual tariff decreases were reduced from 2% to 1%. For offshore wind electricity, the initial tariff was set at € 15¢/kWh (USD 17¢/kWh) until 2015, before decreasing to € 13¢/kWh (USD 15¢/kWh) and reduced by 5% annually thereafter. A "repowering bonus" was also offered to incentivize the replacement of old onshore and offshore turbines as the efficiency was substantially less with the older turbines than modern turbines due to technological advances in wind turbine technology. A € 0.5¢/kWh (USD 1¢/kWh) premium was offered to operators of turbines at least 10 years old on top of whatever their fixed remuneration rate was at the time to replace them with new turbines that were from twofold up to fivefold more efficient than the old turbines.

Tariffs for RE from biomass, geothermal, and hydropower sources were all increased in 2009 *EEG* amendments; however, for solar PV, tariffs decreased for all capacity sizes. By 2009, Chinese solar PV module manufacturers had made tremendous gains in efficiency, and the price per MW cost of solar PV had dropped dramatically. As Chinese exports of solar PV modules began to flood markets around the world – particularly those with attractive FIT programs in place for solar PV – a MET market assessment of the solar PV industry concluded that solar PV rates could be progressively decreased without stifling the growth of solar PV installations. Roof-mounted facility tariffs were decreased to € 43.01¢/kWh (USD 49¢/kWh) up to 30 kW, € 40.91¢/kWh (USD 47¢/kWh) from 30 to 100 kW, € 39.58¢/kWh (USD 45¢/kWh) from 100 kW to 1 MW, and € 33¢/kWh (USD 38¢/kWh) over 1 MW. For free-standing facilities, the tariff was decreased to € 31.94¢/kWh (USD 36¢/kWh). Collectively, the 2009 *EEG* amendments exemplify a sustained commitment by the German government to renewable energy development and realization of the different paces at which different renewable energy technologies were advancing with respect to their capital costs, technological efficiencies, and ease of integration with the existing German electricity transmission grid.

In 2012 the EEG was amended again, and these changes legislatively enshrined stated policy goals for minimum RE electricity generation of 35% by 2020, 50% by 2030, 65% by 2040, and 80% by 2050 (IRENA 2015h). A collection of tariff adjustments also accompanied the 2012 EEG amendments. The tariff for new onshore wind projects had decreased 1% annually as part of the 2009 EEG amendments from € 9.2¢/kWh (USD 11¢/kWh) for the first 5 years of operation to € 8.93¢/kWh (USD 10¢/kWh) in 2012. The 2012 EEG amendment increased the rate of annual premium decrease for new onshore wind installations to 1.5%. The repowering bonus offered under the 2009 EEG amendments was also restricted to turbines put into operation prior to 2002. The fixed € 15¢/kWh (USD 17¢/kWh) tariff for new offshore wind projects was extended to the end of 2018 (instead of 2015 under the 2009 EEG amendments) and then set to decrease by 7% annually (instead of 5%).

An alternative FIT payment model was also offered to offshore wind energy developers in order to accelerate capital cost recovery. While most wind projects were given a 12-year FIT premium guarantee compared to the 20-year guarantees offered for other RE technologies offshore wind developers were offered the opportunity to receive tariff payments of € 19¢/kWh (USD 22¢/kWh) for the first 8 years of project operation instead of the standard offer € 15¢/kWh (USD 17¢/kWh) for 12 years. In addition, the German government-owned development bank KfW began offering a dedicated loan program solely to offshore wind energy developers (KfW n.d.), highlighting the emphasis the German legislature had placed on developing offshore wind energy resources.

In order to contain the escalating costs of solar PV tariff payments and issues with connection of a scattering of small installations to the grid, the 2012 EEG amendment provided for an immediate tariff decrease of 15% for all project sizes as well as annual decreases in the initially offered premiums between 1.5% for projects less than 100 kW and 24% for larger projects. Eligible rooftop facility projects were also capped at no more than 10 MW. Opportunistic developers had rushed to develop solar projects since the 2009 EEG amendments (particularly large-scale projects), and many of the 2012 EEG amendments to the solar power tariff structure were aimed at directly addressing issues that had arisen and public outcry over the increasing cost of electricity. A subsequent EEG amendment called PV-Noelle was introduced in June 2012 to cap total PV solar energy development at 52 GW, after which new solar installations of any size would no longer receive any premium on electricity generated (Jankowska 2014).

In August 2014, the EEG was once again amended as the RE industry in Germany continued to mature. By year-end 2014, electricity from RE sources was projected to supply almost 30% of the net German electricity demand (Berger 2014), and national RE electricity generation goals were increased accordingly to 40–45% by 2025, 55–60% by 2035, and maintained at 80% by 2050. Annual development targets for solar PV at 2.5 GW/year, onshore wind at 2.5 GW/year *net* (i.e., including replacement of old turbines), offshore wind at 6.5 GW by 2020 (equal to approximately 800 MW/year), and biomass at 100 MW/year have been created. FIT premiums offered to developers of new RE projects would be adjusted annually based on the actual pace of development of RE sources compared to these targets. Existing projects are all still paid the premium prices offered under production licenses for a 20-year duration beginning at the start of a project's operational life, subject to the slightly different tariff structures offered for wind energy developers.

Another major change prompted by the 2014 EEG amendment was the introduction of direct marketing requirements for RE electricity generators. Under EU law, all utilities in member countries must be "unbundled," meaning that transmission grid operators cannot own power plants and sell electricity (Bundesnetzagenter 2015a). This was done to prevent vertically integrated utilities from manipulating spot market power prices during periods of peak grid demand by opportunistically shutting down generation at certain facilities in order to collect higher prices from other generation facilities under their ownership. Under the previous FIT arrangement, grid operators were required to purchase all RE electricity produced from RE generation facilities at the premium price agreed to under each facility's production license and then coordinate the sale of this electricity on the electricity spot market. These premiums were recoverable by grid operators from electricity consumers in the form of a renewable energy surcharge that was calculated based on the net premiums paid to RE electricity generators. The net premiums paid to RE electricity producers were calculated quarterly under the ex-post quota system based on the total payments to RE producers minus what the equivalent amount of RE electricity purchased would have cost at the average kWh cost over the quarter. What this allowed RE generators to do was "produce and forget" in other words generate electricity without regard for real-time grid demand, which ultimately placed a significant burden on grid operators to creatively manage the electricity supply and the demand throughout peak and off-peak periods of a day (Energiewende Germany 2015). Under the new direct marketing requirements, all solar PV and wind power operations larger than 100 kW would by 2016 be required to essentially sell their power themselves, either to consumers looking to purchase green electricity under a contracted price or directly on the electricity spot market exchange. If sold directly in the electricity spot market, RE generators are able to recover the difference between the spot market price they receive for their electricity and the rates guaranteed under their production licenses from a renewable energy surcharge paid by consumers, a calculation via a similar method to that previously used to reconcile the additional costs incurred by the grid operators in directly purchasing RE electricity. While the direct marketing requirements may impose a short-term burden on smaller RE power producers unfamiliar with the German power exchange, the long-term goal is to familiarize these operators with electricity markets such that when the 20-year period of guaranteed rates under their original production licenses end, these operators can seamlessly transition to selling into competitive power markets.

By the end of 2014, Germany achieved 39.2 GW of built wind power capacity. A record 5.3 GW of new capacity was added in 2015. Some have speculated this increase in growth may be more due to the fact that FIT rates for offshore wind energy development are set to expire at the end of 2018 and wind developers are therefore rushing to complete projects to lock-in FIT rates; however, the total installed capacity to date still ranks Germany third in the world behind only China and the USA in terms of total installed wind generation capacity. With the end of the offshore wind FIT rates in 2018 and escalating depreciation of the rates offered for onshore wind, wind energy projects will, like solar PV, face increasing competitive pressures in future years as FIT rates decline or expired. As of June 2015, Germany achieved a total of 38.8 GW built PV solar capacity, creating the largest solar PV generating capacity of any country in the world (Bundesnetzagenter 2015b). Given the solar PV development target of 2.5 GW of new solar PV capacity per year and the PV Novelle cap at 52 GW, Germany's FIT rates for new solar PV projects will likely continue until the year 2020, after which all new solar PV projects will be subject to the full pressures of spot market electricity prices.

Germany and Denmark

The German FIT system has influenced the Danish RE industry including wind development on islands such as Samso Island discussed earlier in the chapter. Starting in the mid-1970s, Denmark invested heavily in wind power generation and has been a world leader in the field ever since (DWIA n.d.). In 2014, wind power supplied 39% of all electricity consumed in Denmark (DWIA 2015), and Danish wind power expertise has become a key economic driver in the country, prompting the creation of companies such as Vestas Wind Systems – the largest wind turbine manufacturer and servicer in the world as of 2015 (Vestas 2015) – as well as numerous other wind power service sector companies. The Danish experience with RE closely mirrors that of its larger neighbor to the immediate south, Germany, a country that is widely regarded as one of the most innovative jurisdictions in the world with regard to RE development.

Beginning in 1989 with amendments to the *Electricity Supply Act,* similar to Germany's 1991 *Feed-in Tariff Law,* the Danish Government mandated power suppliers to connect RE projects to the grid and provide an ROFP. The coastal regions of Denmark have experienced significant development of wind energy resources. The timing and substantive content of Danish legislative and policy amendments with respect to RE development is similar to legislative and policy developments in Germany, mostly out of necessity as the Danish-German border forms the only current transmission linkage to mainland Europe for Danish power producers. However, as grid operators in both countries attempt to balance electricity supply and demand, issues have arisen with regard to Denmark's desire to export surplus electricity from its substantive wind energy resources (Starn and Zha 2015).

Citizens in the northernmost German province of Bavaria have opposed new transmission lines that would be required to carry additional electricity from Denmark and other Nordic countries to consumers in southern Germany and other European Union countries. Periodically throughout the year, the spot price for electricity on the Nordic electricity exchange, or Nord Spot Pool, a centralized electricity exchange for all electricity producers in Denmark, Finland, Sweden, Norway, Estonia, and Lithuania (Nord Spot Pool 2015), has declined to negative values due to oversupply from wind energy sources at peak times of certain days when the wind is stronger in particular regions. This means that all Danish electricity generators operating during these times are essentially *paying* domestic consumers to consume electricity during these peak periods of oversupply. In the absence of new grid connections with Germany's mainland power grid, Danish energy producers will continue to be forced to sell into an oversupplied market within the Nord Spot Pool. While wind energy producers in Denmark still receive an "environmental premium" of DKK 0.10/kWh (CAD 2¢/kWh) for every kWh produced which helps to compensate for the periods when the market pool price is negative, Denmark serves as one cautionary example of one country's RE ambitions outpacing that of other jurisdictions whose cooperation is required for a profitable outcome. While Denmark's wind energy industry has undoubtedly flourished in recent decades thanks to its early start globally in wind energy development and a variety of domestic incentive programs, its collective success is capped based on the degree of cooperation from its southern neighbors. In light of this transmission bottleneck, Denmark has begun to explore direct transmission grid linkages with other European countries, with an underwater 700 MW line to the Netherlands slated to be operational by 2019. Other Nord Spot Pool members have also begun to explore solutions to the electricity oversupply problem currently facing member countries. For example, Norway and Britain have an agreement in principle to construct the world's largest subsea transmission line – a 1400 MW line that would connect the two countries and provide another outlet for Nord Spot Pool member electricity during periods of oversupply.

Ecuador

In 2000, Ecuador became one of the first countries in South America to adopt a feed-in tariff (FIT) model to promote the development of renewable energy. While a number of setbacks regarding the application and enforcement of the program led the government of Ecuador to suspend the program in 2009, the Ministry of Electricity and Renewable Energy (MEER) reenacted in 2011 by Decree 004/11 (CONELEC Regulación No. 004/11) a revamped FIT program with added provisions based on lessons learned from the FIT program that ran from 2000 to 2009. The new FIT program has offered a variety of tariffs for electricity generated from wind, solar PV, biomass, geothermal, and hydropower sources on the Ecuador mainland, as well as a separate set of rates for power generated from these sources on the isolated Galapagos Islands designed to reduce reliance on imported diesel fuel for electricity generation. An open call for tenders was issued to RE developers that provided for 15 year power purchase agreements (PPAs) to be issued to successful bidders. Ecuador's tremendous solar energy potential, combined with the newly offered FIT solar PV rate offering of USD 40.03¢/kWh, has prompted a significant increase in the number of solar energy companies bidding for PPAs to supply electricity to the grid. When the original call for tenders expired December 31, 2012, the Ecuadorian electricity regulator CONELEC had agreed to PPAs with 17 companies for 288 MW of combined solar PV capacity (Lopez 2015).

In June 2013, Decree 001/13 (CONELEC Regulación No. 001/13) replaced the 2011 FIT program that had expired in December 31, 2012 with minor amendments. Tariffs for energy from wind, biomass, geothermal, and hydropower sources for both continental Ecuador and the isolated Galapagos Islands were adjusted slightly, and, most significantly, the tariffs for solar PV were eliminated and tariffs for solar thermal offered instead. With 288 MW of planned solar PV capacity already in place from the 2011 call for bids, rate offerings were changed to incentivize development of other forms of RE such as solar thermal and wind. CONELEC has issued a call for bids to renewable energy developers until December 31, 2016, or until a target 6% of the national installed electricity capacity is achieved. All electricity generated from RE sources will be given priority purchase on the grid under 15 year PPAs, similar to the provisions offered to RE developers under the 2011 call for bids. While development of many RE projects particularly solar PV has been slower than expected in Ecuador, by the end of 2014, almost 27 MW of solar PV capacity had been installed, from only 4 MW of installed capacity at the end of 2013, representing an almost 600% increase in solar PV capacity in the country. Ecuador now has the second-largest solar PV capacity in South America, trailing only Chile's installed capacity by year-end 2014. While some of the original projects planned as part of the awarded 288 MW of capacity have been either temporarily or permanently suspended for a variety of economic, regulatory, and social concerns, Ecuador's solar PV capacity will continue to undergo exponential growth as more of the projects become operational.

Renewable Portfolio Standards (RPS)

Another common approach used by governments to promote RE development has been the creation of renewable portfolio standards (RPS). In a RPS system, the government usually sets renewable electricity generation goals as a certain percentage of annual electricity generation to be achieved by a certain date. A 3–5-year grace period between the date of the announcement and the first goal threshold to be achieved to allow utilities in the jurisdiction to either individually develop renewable energy projects to add to their electricity portfolio mix or to purchase renewable electricity certificates (RECs) from third-party project developers that can be claimed as credit equivalents toward their RPS requirements. After the grace period ends, the RPS target for each utility increases annually or every 3–5 years, although these increases are frequently subject to future changes based on the cost of electricity in some jurisdictions, the feasibility of utilities attaining RPS targets, or the availability of RECs for purchase. A utility that fails to meet their RPS requirements is often subject to heavy fines or other fiscal consequences. RPS systems are common in jurisdictions with private utilities and liberalized electricity markets or power trading pools. Japan, the USA, and Chile provide examples of countries that have adopted RPS systems to promote RE development.

Japan a nation comprised of large and small islands lacks domestic fossil fuel resources and is therefore more susceptible to global oil price fluctuations. During the Arab Oil Embargo of 1973 and oil price shock following the Iranian Revolution in 1979, Japanese consumers and industry experienced significantly higher prices for electricity, fuel, and other energy resources. The lasting impact of these oil crises prompted the Japanese government to adopt in 1980 the *Law Concerning Promotion of Development and Introduction of Oil Alternative Energy*, which provided for the establishment of the New Energy Development Agency (NEDO) (IRENA 2015k). NEDO assumed the responsibility to coordinate research and development on RE technologies such as solar PV and wind. Japan grew to be the world's second-largest producer of solar PV electricity by the early 2000s under the supervision of NEDO largely thanks to a government subsidy program for household rooftop installations. Then in April 2003, the government adopted a *Special Measures Law Concerning the Use of New Energy by Electric Utilities* (SML) (Mendona 2007). The law creates an RPS system for all electricity wholesalers that mandates an increasing annual percentage of electricity generation from wind, solar, geothermal, biomass, or small-scale hydropower sources based on a 2010 target of 12.2 billion kWh of electricity from RE sources or approximately 1.35% of the total estimated electricity demand by 2010. While seemingly low compared to more recent RPS requirements imposed in other countries on electricity wholesaler and utilities, the Japanese government did not want to hinder the productivity of its large industrial manufacturing base through electricity rate increases that often accompany the installation of new RE generation capacity. Additionally, Japan at the time was the third-largest electricity producer in the world after the USA and China, and the 1.35% RPS target by 2010 was seen as generally being in line with the RE generation targets that the USA and China had set at the time for themselves (BP 2015).

The RPS imposed under the SML has established flexible dates for satisfying RPS requirements by back-loading RE generation targets to the end of the 2003–2010 period for Japan's larger utilities in order to allow time for construction of new RE generation capacity (Nakakuki and Kudo 2003). Electricity wholesalers had two main options available to meet their RPS requirements: build sufficient RE generation capacity themselves or purchase an equivalent amount of electricity derived from RE sources via an REC-style system. Alternatively, a financial penalty would be imposed should wholesalers fail to meet their RPS requirements. The RPS system, combined with the rising prices of oil and natural gas throughout the 2000s, prompted installed RE generation capacity at an average annual rate of 8% from 2003 to 2009 (Edahiro 2014). In 2009, the Japanese government passed a further *New Purchase System for Solar Power-Generated Electricity* regulation, that mandated utilities to buy all excess electricity produced from rooftop solar PV at fixed prices – 48 JPY/kWh (USD $0.39/kWh) for electricity from domestic household rooftops and JPY 24/kWh (USD $0.20/kWh) from non-household rooftops (IRENA 2015l). The cost of this FIT-style program was recovered by utilities through a JPY 30/month (USD $0.24) surcharge on all grid-connected electricity consumers and was set to remain open for a period of 10 years.

The combined outcome of the SML and additional programs such as that of the *New Purchase System for Solar Power-Generated Electricity* regulation facilitated the island nation exceeding its RPS targets by 2010. In 2012, the Japanese government introduced the *Act on Purchase of Renewable Energy Sourced Electricity by Electric Utilities* that created the island nation's first FIT program for electricity from RE sources (IRENA 2015m). The FIT program adopted in 2012 not only replaced the RPS system in place from 2003 to 2010 but also the existing FIT program for solar PV that had been in place since 2009 with new rates. Premium rates for electricity from rooftop solar PV, wind, geothermal, hydropower, and a variety of biomass sources were offered to prospective RE developers along with 10–20-year PPAs depending on the RE source and technology (IRENA 2015m). The Ministry of Economy, Trade, and Industry (METI) annually reviewed the offered rates as development of different RE sources progressed or technological and manufacturing advances made certain technologies, especially solar PV, more affordable.

The JPY 42/kWh (USD $0.34/kWh) rate for both residential (<10 kW) and nonresidential (>10 kW) rooftop solar PV under the original offer period was decreased approximately 10% by METI to JPY 38 (USD $0.31/kWh) and JPY 37.8 (USD $0.31/kWh) in April 2013 after a massive influx of applications for solar PV project grid connections. From the start of the program in 2012 to the 2013 year-end, Japan's total installed solar PV capacity had more than doubled, from approximately 6.6 GW to 13.6 GW (IEA 2014). In April 2014, METI further reduced the rates for residential and nonresidential solar PV to JPY 37 (USD $0.30) and JPY 32 (USD $0.26), respectively, as the costs for solar PV panels continued to drop dramatically and applications for rooftop solar PV connection continued to increase. By the 2014 year-end, Japanese solar PV capacity had grown by another 9.7 to 23.3 GW of total capacity, making Japan the third-largest solar PV electricity producer in the world after Germany and China (IEA 2015). Despite the restart of designated nuclear reactors in September 2015, Japan has experienced explosive growth in its RE portfolio since introduction of the FIT program in 2011 under the *Act on Purchase of Renewable Energy Sourced Electricity by Electric Utilities*. In response to the Fukushima Daiichi disaster, a shift in Japanese energy policy toward RE over nuclear is illustrated by the world's largest offshore wind farm currently being planned off the Fukushima coast. The project is to have a collective 1 GW capacity once complete by using massive floating turbines between 2 and 7 MW in capacity (Yirka 2013, Watanabe 2015). Even with nuclear power slowly returning to Japan's grid, plans to continue the FIT program subject to an annual review seem set to continue for several years to come. Japan currently produces approximately 10% of its electricity from RE sources, with a 20% RE generation target by 2020 (Harlan 2013). While it remains to be seen if the RE target be achieved under the existing FIT program alone, collectively, the island of Japan provides another example of a country prompted to develop political and technological ingenuity in the face of scarce fossil fuel resources.

The most common approach used by state governments in the USA to promote RE development has been to create renewable portfolio standards (RPSs), as they are flexible market-based policies created to ensure that public benefits of renewable energy are recognized (IRENA 2015b). An RPS requires electricity providers within the state to obtain a minimum percentage of their power from renewable energy resources by a certain date. Each state/regional government chooses to fulfill its RE mandate by using a several renewable energy sources, including wind, solar, biomass, geothermal, or other renewable sources depending upon the availability of the different RE sources. Some RPSs will specify the technology mix, while others leave it up to the market. Currently there are 33 states in the USA plus the District of Columbia that have RPS requirements or goals in place. While the first RPS was established in 1983, the majority of states adopted or strengthened their standards after 2000. A central component of an RPS in the USA is the RE requirement that RE development be implemented through a system of tradable renewable electricity credits (RECs). Retail electricity sellers can satisfy the RPS requirement by either generating renewable electricity themselves or by purchasing RECs from other generators.

Chile

Chile provides another example of a nation that has adopted a RPS system to promote solar energy development. The country has an extensive coastline, and desert lands in northern Chile offer some of the best conditions for solar power development anywhere in the world. The Chilean government originally opened the door to small-scale renewable energy development in 2004 through the introduction of *Short Law I* (Law 19.940). This law enabled any electricity generator with the right to sell electricity into a deregulated market and exempted generators under 20 MW from transmission tariffs normally required to be paid by a generator for the right to be connected to the transmission grid. Subsequent to *Short Law I*, the *NCRE* Law (Law 20.257) was adopted in 2008 and was specifically tailored to increase the share of electricity on the Chilean grid generated from RE sources.

Non-Conventional Renewable Energy Law (Law 20.257) (2008)

The national government has been active in promoting development of northern Chile's solar resources by offering financial security to international solar developers via loan guarantees and subsidies for projects under a Support for Non-Conventional Renewable Energy Development Program (IRENA 2015c). In April 2008, Chile enacted Law 20.257, otherwise known as the *Non-Conventional Renewable Energy (NCRE) Law*, created to promote the development of RE sources such as solar, wind, and geothermal by setting a minimum renewable portfolio standard (RPS) for all electricity producers in the country of 5% by 2014, increasing 0.5% annually (IRENA 2015d). The RPS originally came into force in 2010 and has since been amended to require all large electric utilities (greater than 200 MW operational capacity) to achieve a 20% RPS by 2025, equal to a 1.5% increase annually from 2015 to 2025 (IRENA 2015d).

Electricity generators are to satisfy the RPS requirements through either construction of their own facilities or by contracting with third parties for the purchase of their production. The *NCRE Law* has capped hydropower generation in Chile that can be counted toward utilities' RPS requirements for projects no more than 20 MW in net capacity; however, hydropower generating stations that are between 20 MW and 40 MW may still count the first 20 MW generation block toward RPS requirements. With approximately 33% of electricity already generated in Chile from renewable hydropower sources, this *NCRE Law* provision forces utilities to pursue development of a broader mix of RE sources such as solar and wind that contributes to a more diversified RE generation portfolio. If utilities are unable to meet their RPS target, they are fined 16,176 CLP/MW/year ($31.90 CAD/MW/year) for the difference between their realized annual renewable electricity generation and that required under their RPS requirements for that year.

Despite the efforts by the Chilean government in recent years to promote renewable energy development, it is anticipated that the country will not achieve the RPS targets set under the *NCRE Law*. Utilities will instead simply pay the fines for not meeting their targets. Despite the creation of a new Ministry of Energy in 2010 and dedication of considerable resources to attracting renewable energy developers to Chile, a number of technical, financial, and institutional barriers still face prospective developers, notwithstanding the reforms brought under the *Short Law I* and the *NCRE* Law (von Hatzfeldt 2013). While almost 9% of Chile's electricity generation in 2014 was from renewable energy sources, this is still significantly less than the 20% by 2025 target set by *NCRE Law* amendments in 2013. In light of this, the Chilean government has considered a call for tenders from renewable energy developers for a series of PPAs to entice renewable energy developers with guaranteed rates of return (Nielsen 2013). Banks and other financial institutions are often far more willing to loan money at lower interest rates for projects with a guaranteed rate of return rather than those at the mercy of a market as in the deregulated spot electricity market in Chile. While Chile is set to become the largest solar power producer in Latin America, with over 400 MW of solar capacity set to be operational by year-end 2016, it remains unclear as to how the Chilean government plans to meet the balance of its target renewable energy capacity by the original target dates (Nielsen 2013).

Periodic Call for Tenders

A third basic approach used by governments to finance RE development is a periodic call for tenders from RE developers. A call for tenders is used in jurisdictions with large state-owned utilities. Rather than guarantying a specific price per kWh for all developers and grid access once operational as under a FIT program, or legislating utilities to supply or purchase from a third party a certain percentage of their electrical generation through an RPS system, periodic auctions or a call for bids from RE developers is used often in jurisdictions with large public utilities. Governments establish a specific generation capacity or portfolio mix it intends to build and then proceed to choose from invited proposals from proponents, whose RE projects will be awarded guaranteed purchase and grid connection agreements, usually in some form of a PPA contract similar to that issued under a FIT program. A government will usually begin by issuing a request for an expression of interest (RFEOI) from RE developers to garner industry support in developing certain RE sources within their jurisdiction. Should the RFEOI yield sufficient interest, the government may proceed to issue a request for proposals (RFPs) from developers which typically have a host of details for content to be submitted in a proposal and the evaluation criteria that the determining body plans to use in selecting projects. Successful applicants are typically awarded some form of a power purchase agreement (PPA) that guarantees a certain price for electricity output from the project for a specific number of years. The price is usually set on a case-by-case basis after negotiations between the project proponent and government or public utility. One major advantage of the call for tender process is that a government or public utility is able to defer significant costs associated with upstream project planning. As well, while a government or public utility may provide an outline of what they are looking for in terms of total generating capacity, project size, and type by issuing an RFEOI or RFP, a privilege clause may be used that provides the issuer with significant discretion in terms of project selection or even whether to select any projects at all.[24]

---

[24] For a broad summary of "privilege clause" utility in Canada, for example, see M.J.B. Enterprises Ltd. *v.* Defense Construction (1951) Ltd., [1999] 1 S.C.R. 619.

One Other Challenge to RE Development for Some Nations: International Trade Agreement Obligations Under GATT

In 2006, Ontario was the first Canadian province to adopt a Renewable Energy Standard Offer Program (RESOP), which ultimately paved the way for its feed-in tariff (FIT) program in 2009. Ontario has promoted RE development through a combination of FITT and RPS initiatives. The Ontario FIT program has prompted wind development on Wolfe Island and in other areas along the shoreline along with solar development elsewhere in the province. In 2006 the Ontario Power Authority (OPA) launched the Renewable Energy Standard Offer Program (RESOP) to make it easier for small RE generating facilities to participate in the electricity supply system through local electricity distributors (IRENA 2015a). Developers of solar, wind, additional hydropower (waterpower under the RESOP), and biomass projects from 1 kW to 10 MW in capacity were eligible for a fixed per kWh rates – CAD 42¢/kWh for solar photovoltaic (PV) systems and from CAD 3.52¢/kWh to CAD 11.08¢/kWh for wind, waterpower, and biomass, depending on peak or off-peak grid hours. The rates were guaranteed to developers of RE projects by 20-year contracts with the OPA. As North America's first true FIT program, the RESOP aimed to generate 1000 MW of new RE generation capacity over a period of 10 years. In a little more than one year, however, contracts for RE projects representing over 1300 MW of new generation capacity had already been signed with the OPA. By January 2009, this figure had reached 604 contracts totaling 1518 MW of new capacity. By RE technology source, RESOP has supported 106 wind energy projects totaling 813 MW, 434 solar PV projects totaling 527 MW, 33 bioenergy projects totaling 99.5 MW, and 31 hydro projects totaling 78 MW. After a review by the Ontario government in 2009, the RESOP was extended to the Ontario feed-in tariff (FIT) program with a new FIT regime for a variety of RE technologies and expanded eligibility for microgeneration projects.

In 2009, *The Green Energy and Economy Act* (or *Green Energy Act*) (GEA) was adopted in Ontario, Canada, to further promote RE development under a provincial FIT Program (Bosworth 2010). Premiums were paid for electricity generated from various RE sources, including rooftop and ground-mounted solar PV, wind, waterpower, biomass, on-farm and off-farm biogas, and landfill gas. Successful project applicants were awarded PPAs that guaranteed the offered FIT Program premiums for electricity generated from these projects for a period of 20 years once operational and contained a domestic content manufacturing requirement (DCR) that between 50% and 60% of RE project goods and services by cost be sourced from Ontario manufacturers and service providers (Timmins and Kirsh 2012). Initial premiums ranged from CAD 10.3¢/kWh for landfill gas projects larger than 10 MW to 80.2 CAD¢/kWh for residential solar rooftop projects smaller than 10 kW, with rates set to be reviewed annually by the Ontario Independent Electricity System Operator (IESO) based on electricity grid demand and transmission planning requirements, changing project development costs for different RE technologies, and ongoing monitoring of project applications from developers to ensure a balanced RE generation portfolio mix from different sources. The FIT Program also provided specific incentives in the form of a "price-adder" ranging from CAD 0.5 to 1.5¢/kWh on top of FIT Program rate premiums for aboriginal- and community-based projects in which the local group claims at least a 15% equity share in the project. The price-adder mechanism was meant to increase efforts by RE project developers to consult with communities in which their projects are located and attain local buy-in or a social license to sometimes controversial projects (e.g., wind). Even with the price-adder mechanism, however, Ontario has experienced a substantial increase in litigation related to RE projects since enactment of the *GEA*, especially in regards to large wind energy projects. Opponents have brought forth a variety of legal arguments in regards to these projects, mostly revolving around some form of perceived negative health and safety effects, visual impacts on local property and decreases in land values, encroachment on endangered and/or at-risk species habitat, threats to bird or bat populations, or just general opposition to wind energy development in certain regions – otherwise known as "not-in-my-backyard" (NIMBY).[25]

Soon after the adoption of the GEA in Ontario, the provincial government began facing increasing scrutiny after a surge of applications from RE project proponents interested in pursuing highly attractive FIT rates, particularly those developing solar PV. The new PPAs awarded under the FIT Program to large solar and wind farm developers were being blamed by industry and consumer groups for escalating electricity prices in Ontario. Much debate existed about the degree to which FIT Program PPAs had increased the cost paid by electricity consumers in Ontario (OEB 2012; Bridgepoint Group 2012, ECO 2011), and the GEA became a key election issue in the 2011 Ontario provincial election, with the provincial Progressive Conservative (PC) Party campaigning on a platform that included eliminating the GEA and the FIT Programs (Howlett and Ladurantaye 2011). A 2 year review of the FIT Programs was announced and culminated in the reduction of rates for rooftop and ground-mounted solar PV of various sizes from 10% up to 32% for solar PV projects <10 kW and for wind of all sizes by 15%. Tariffs for all other RE technologies remained the same. In 2013, a 2 year review of the FIT Programs brought a renewed focus on small and medium scale projects via the elimination of FIT Program rates for projects larger than 500 kW in capacity (Timmins and Blumer 2013). Developers of RE projects larger than 500 kW would now be subject to a competitive procurement process in which the OPA will make periodic calls for tenders to RE developers based on recommendations made by the IESO. Consideration of applications for FIT-eligible RE projects between 10 and 500 kW and all microFIT projects would continue as before, although a 900 MW target of additional RE capacity by the end of 2018 was set, after which applications for these projects will presumably be restricted as well. Under the 2 year RE program review in 2013, rates for electricity from solar and wind sources were also further decreased, while rates for electricity

---

[25] See, for example, Hanna v Ontario (Attorney General), 2010 ONSC 2660; or Kenney v Municipal Property Assessment Corp., Region No. 05, 2012 CarswellOnt 3747; or Drennan v Director, Ministry of the Environment, 2014 CarswellOnt 1695.

from waterpower, biomass, biogas, and landfill gas were all increased in order to stimulate their development of these RE resources. The 2015 2-year review process remains ongoing, with results and recommendations due in early 2016 (IESO 2015).

In addition to the debate surrounding the escalating costs to electricity customers, government policies and laws to promote RE development in Ontario prompted an international trade dispute after Japan brought a challenge before the (WTO) in September 2010 on the basis that the DCR of the *GEA* discriminated against foreign firms in the RE industry contravening the General Agreement on Tariffs and Trade (GATT) 1994 and the Agreement on Trade-Related Investment Measures (TRIMs). In 2011, the European Union also joined Japan in challenging the *GEA* DCR provisions on behalf of member states such as Denmark, Germany, and Spain (with large RE industries). As an original signatory of the GATT 1947 and a founding member of the WTO in 1995, Canada is bound by the general trade rules of the GATT 1994 (the modern day successor to GATT 1947) and legally bound to abide by decisions from WTO tribunals concerning its domestic industries and trade practices. While GATT 1994 forms the umbrella treaty governing WTO oversight of the international trade of goods, the Agreement on TRIMs concluded at the end of the Uruguay Round of GATT negotiations in 1994 also is a key element of WTO oversight. The Agreement on TRIMs prohibits any TRIMs that are inconsistent with Articles III or XI of GATT 1994 and serves to restrict certain investment measures that distort the trade of goods. In effect, the Agreement on TRIMs is a set of rules that prevents WTO member countries from favoring domestic firms over international competitors within domestic industries covered by the GATT 1994.

When first adopted in 2009 by the Ontario government, the Green Energy Act provided that all RE projects under the provincial FIT Program have between 50% and 60% of project goods and services by cost be sourced from Ontario-based manufacturers and service providers (Timmins and Kirsh 2012). While the purported aim of the domestic content provisions was to stimulate the creation of 'green" jobs and employment in Ontario – with a particular focus on the replacement of jobs in the declining manufacturing sector of southwestern Ontario – the domestic content provisions unfortunately contravened Canada's trade obligations under GATT 1994 and the Agreement on TRIMS.

Under the Canadian constitution as the national government has jurisdiction over international trade matters, the federal government represented the regional Ontario government in the dispute and argued that the Domestic Content Requirements (DCR) under the FIT Program were not subject Article III:4 of GATT 1994. The basis of this argument was that because the primary aim of the FIT Program was to secure a clean RE supply of electricity for Ontario consumers and not for the international commercial sale of RE equipment, and the program qualified under Article III:8(a) of GATT, which excuses national treatment obligations with regard to procurement by governmental agencies of products for governmental use. The electricity industry in Ontario as in many other jurisdictions is a monopoly. However, in May 2013, a World Trade Organization (WTO) tribunal ultimately ruled against Canada in favor of Japan and EU in concluding the DCR for investment in the production of RE technology in Ontario and that contracts under the FIT program, by favoring the use of domestic over imported products, are in fact "trade-related" and therefore not exempt under Article III:8(a) (Kanargelidis et al. 2013; WTO 2012). As a result, the WTO found Canada in violation of Article III:4 of GATT and Article 2.1 of the Agreement on TRIMs. In June 2013, the Ontario government announced that it would comply with the WTO ruling to avoid any potential trade sanctions against Canada as a whole and removed all DCR provisions from new FIT and microFIT Program contracts. While the DCR provisions in the FIT Program attempt to promote RE development by cultivating a domestic RE manufacturing and service industry in Ontario, the recent trade dispute serves as an important reminder of how RE policies and law cannot be developed in a vacuum and are subject to existing national and international trade agreements frameworks and treaty obligations.

Conclusion

Historically on islands that lack domestic fossil fuel reserves such as Japan and Iceland, governments have created policies and adopted laws to promote RE development to provide national energy security. In this chapter we have examined how government policies and laws have played an important role in facilitating RE development. Geothermal and wind provide examples of two important RE resources that have a proven track record of electricity production on islands. As has been the case with other sources of energy, the potential for developing RE on islands depends upon the availability and cost of competing sources of energy such as fossil fuels and the availability and intensity of the specific RE resource itself such as wind that will vary according to the geographic position and geology of the island.

As with larger island nations, economic challenges faced by governments of the Small Island Developing States include higher diesel costs due in part to higher transportation costs. Recently solar energy development on small remote islands has increased. In addition to securing financing for solar RE projects on remote islands, one additional challenge on islands subject to hurricanes is the damage to roof tops and solar PV panels from strong winds. The availability of financing for solar projects either from national governments or through international assistance can be an important factor in whether solar development proceeds on a remote island. The potential for hurricane damage should be an important planning consideration for the placement of panels to minimize damage in hurricane prone areas. As water is required to clean the solar panels for optimal electricity generation, water usage is also a consideration for solar projects on islands with limited water supplies. In addition to electricity generation, solar energy has been used for many years in solar hot water systems, a second benefit from developing this RE resource.

As nonsaline water is a limited and precious resource on numerous remote islands, the energy-water nexus should be considered when planning for solar energy and geothermal projects. One important consideration in wind and solar development is the variability in generating capacity and the availability of alternate energy sources that can be brought online to maintain a reliable electricity supply for customers. Unlike solar and wind, geothermal energy has the advantage of providing ongoing electricity generating capacity which in part may explain the extensive development of geothermal energy on volcanic islands where the geothermal resource is available.

Geothermal energy is one of the RE resources that has been developed on numerous islands for several decades. Favorable island geology which includes volcanic rocks is an important factor in determining whether economic development of the geothermal resource is viable. The Icelandic geothermal energy policy and regulatory framework has provided a useful template for other island governments interested in developing their geothermal energy resources. More than four decades ago, to attract investment in geothermal energy development on the remote island, the government of Iceland has created certainty in the legal framework surrounding geothermal rights to provide to prospective geothermal developers with sufficient incentive to make the significant capital investment in a geothermal power plant. In Iceland another important factor that has contributed to the success of geothermal development on the island is a supportive government policy and an associated legal framework that assigns geothermal energy ownership rights to surface land owners in the same manner that subsurface mineral rights are allocated to minimize disputes that could delay and frustrate geothermal development. In the Philippines, geothermal energy has resulted in water contamination at some sites.

With regard to wind energy development, Samso Island in Denmark provides an example of successful wind and biomass energy development, and Ramea Island and Prince Edward Island in Canada provide examples of technological innovation in wind energy technology that has led to successful wind energy projects on small remote islands. Government support for wind energy in both countries has been critical to successful electricity generation. Tidal energy has not been developed to the same level on remote islands as geothermal and wind resources and is currently an active area for experimentation and technological innovation.

While minor differences exist in the countries that we have discussed with respect to policies designed to promote RE development and the laws that provide the framework to implement the policies, the following three general approaches have been used by numerous governments to facilitate RE development at the national level:

1. Feed-in tariff (FIT) programs
2. Renewable portfolio standard (RPS) systems
3. Periodic calls for tenders from renewable energy developers

Germany, Denmark, Japan, the USA, Canada, and Ecuador provide a few examples of countries where governments have adopted feed-in tariff (FIT) programs. Under this type of program, the government typically provides an ongoing open offer to RE developers that RE projects will be connected to the main transmission grid once operational with the cost usually borne by the utilities, not the RE developers. Frequently there is a right of first purchase (ROFP) or priority grid access to ensure that all generated electricity is sold. Often, premiums (i.e., tariffs on the purchasing utility or downstream customer) are offered for electricity from RE projects which guarantees a minimum price per kilowatt-hour (kWh) above the average grid electricity price that varies by RE technology. The FIT rates are often set to correspond to calculated break-even costs for certain RE technologies to assist project developers in recovering upfront capital costs that are often substantially higher than those for electricity producers with plants that operate on fossil fuel-based sources. These premium rates are often offered to individual project developers for a period of 15–20 years that are guaranteed under either a contractually binding power purchase agreement (PPA).

As one of the first actors to adopt a FIT system and the Electricity Feed-in Law of 1991, Germany has a longer history of RE policy experimentation than most countries in the world – particularly with regard to FITs that have become popular in many jurisdictions around the world. Germany is widely regarded as having the most successful RE development track record of any country in the world. During the last 25 years, Germany, a nation with extensive coastal RE development, has been a pioneer in the creation and refinement of FIT programs and has continually revised its RE policy and legislation as technological innovation and experience from the deployment of RE wind and solar technology have provided insight to improve the effectiveness of the FIT program. The German FIT system has influenced the Danish RE industry including wind development on islands such as Samso Island discussed earlier in the chapter. Starting in the mid-1970s, Denmark invested heavily in wind power generation and has been a world leader in the field ever since (DWIA n.d.). In 2014, wind power supplied 39% of all electricity consumed in Denmark (DWIA 2015), and Danish wind power expertise has become a key economic driver in the country, prompting the creation of companies such as Vestas Wind Systems, the largest wind turbine manufacturer and servicer in the world as of 2015 (Vestas 2015), as well as numerous other wind power service sector companies. The Danish experience with RE closely mirrors that of Germany, a country that is widely regarded as one of the most innovative jurisdictions in the world with regard to RE development. As Germany has been very successful in promoting coastal wind energy development and continues to be a world leader in developing and implementing RE, governments and regulators interested in promoting RE development including those that manage RE development on remote islands should reflect on the German RE experience.

A second common approach used by governments to facilitate RE development has been the creation of renewable portfolio standards (RPS). In a RPS system, the government usually sets renewable electricity generation goals as a certain percentage of annual electricity generation to be achieved by a certain date. A 3–5-year grace period between the date of the announcement and the first goal threshold to be achieved to allow utilities in the jurisdiction to either individually develop renewable energy projects to add to their electricity portfolio mix or to purchase renewable electricity certificates (RECs) from third-party project developers that they can be claimed as credit equivalents toward their RPS requirements. After the grace period ends, the RPS target for each utility increases annually or every 3–5 years, although these increases are frequently subject to future changes based on the cost of electricity in some jurisdictions, the feasibility of utilities attaining RPS targets, or the availability of RECs for purchase. A utility that fails to meet their RPS requirements is often subject to heavy fines or other fiscal consequences. RPS systems are common in jurisdictions with private utilities and liberalized electricity markets or power trading pools. Japan, the USA, and Chile provide examples of countries that have adopted RPS systems to promote RE development.

The third basic approach used by governments to finance RE development that we have considered is a periodic call for tenders from RE developers. A call for tenders is used in jurisdictions with large state-owned utilities. Rather than guarantee a specific price per kWh for all developers and grid access once operational as under a FIT program or legislating utilities to supply or purchase from a third party, a certain percentage of their electrical generation through an RPS system, periodic auctions, or a call for bids from RE developers is used often in jurisdictions with large public utilities. Governments establish a specific generation capacity or portfolio mix it intends to build and then proceed to choose from invited proposals from proponents, whose RE projects will be awarded guaranteed purchase and grid connection agreements, usually in some form of a PPA contract similar to that issued under a FIT program. Successful applicants are typically awarded some form of a power purchase agreement (PPA) that guarantees a certain price for electricity output from the project for a specific number of years. The price is usually set on a case-by-case basis after negotiations between the project proponent and government or public utility. One major advantage of the call for tender process is that a government or public utility is able to defer significant costs associated with upstream project planning.

In some countries a combination of the basic approaches discussed above has been used by governments. One other obstacle to RE development that we have considered is the 2012 WTO decision under GATT in which the panel concluded that the domestic content requirements for equipment used in RE projects directed toward increasing RE development in the jurisdiction under consideration were contrary to international trade agreement obligations. The decision undermines the efforts of governments attempting to increase domestic support for RE development by enhancing the benefits of domestic RE development.

References

Act on the Survey and Utilisation of Ground Resources, Statutes of Iceland, Act 57/1998. Article 1. Unofficial translation. http://eng.atvinnuvegaraduneyti.is/media/acts/Act-No-57-1998-on-survey-and-utilisation-of-ground-resources.pdf. Accessed 31 Aug 2015

Atlantic Canada Opportunities Agency [ACOA] (2014, January 15) Harper Government Invests in Innovation in Newfoundland and Labrador. http://www.acoa-apeca.gc.ca/eng/Agency/mediaroom/NewsReleases/Pages/4211.aspx. Accessed 31 Aug 2015

BC Hydro (2015) Acquiring power—clean power call. https://www.bchydro.com/energy-in-bc/acquiring_power/closed_offerings/clean_power_call.html. Accessed 31 Aug 2015

Berger B (2014) Electricity production from solar and wind in Germany in 2014. Fraunhofer Institute for Solar Energy Systems ISE. https://www.ise.fraunhofer.de/en/downloads-englisch/pdf-files-englisch/data-nivc-. Accessed 31 Aug 2015

Bertani R (2015) Geothermal power generation in the world 201-2014 update report. In: Proceedings of the world geothermal congress, Melbourne. https://pangea.stanford.edu/ERE/db/WGC/papers/WGC/2015/01001.pdf. Accessed 31 Aug 2015

Bosworth M (2010) North Star: Ontario guides the way with North America's first true feed-in tariff. Photon Magazine, February 2010 Issue. PV Coast to Coast Series—Part 3: Ontario. http://d-bits.com/cellar/PHOTON_USA_2010-02_Ontario.pdf. Accessed 31 Aug 2015

BP plc [BP] (2015) BP statistical review of world energy—June 2015. http://www.bp.com/en/global/corporate/about-bp/energy-economics/statistical-review-of-world-energy.html. Accessed 31 Aug 2015

Bridgepoint Group (2012) Review of the December 2011 annual report by the office of the auditor general of Ontario – renewable energy facts: Ontarians have a good deal. http://www.bridgepointgroupltd.com/database/rte/files/Renewable%20Energy%20Facts.pdf. Accessed 31 Aug 2015

Bundesnetzagenter (2015a) Electricity and gas—unbundling, Concessions, Complex Networks. http://www.bundesnetzagentur.de/cln_1431/DE/Sachgebiete/ElektrizitaetundGas/Unternehmen_Institutionen/EntflechtungKonzessionenArealnetze/entflechtungkonzessionenarealnetze-node.html. Accessed 31 Aug 2015

Bundesnetzagenter (2015b) Electricity and gas—data messages and EEG remuneration rates for photovoltaic systems. http://www.bundesnetzagentur.de/DE/Sachgebiete/ElektrizitaetundGas/Unternehmen_Institutionen/ErneuerbareEnergien/Photovoltaik/DatenMeldgn_EEG-VergSaetze/DatenMeldgn_EEG-VergSaetze_node.html. Accessed 31 Aug 2015

Cardwell D (2015, January 17) Green-energy inspiration off the coast of Denmark. The New York Times. http://www.nytimes.com/2015/01/18/business/energy-environment/green-energy-inspiration-from-samso-denmark.html. Accessed 31 Aug 2015

CONELEC Regulación No. 001/13, Regulations of Ecuador

CONELEC Regulación No. 004/11, Regulations of Ecuador

Danish Wind Industry Association [DWIA] (2015) The Danish Market: statistics on the development of wind power in Denmark 2005–2014. http://www.windpower.org/en/knowledge/statistics/the_danish_market.html. Accessed e31 Aug 2015

Danish Wind Industry Association [DWIA] (n.d.) Wind energy. Ministry of Foreign Affairs of Denmark. http://denmark.dk/en/green-living/wind-energy/. Accessed 31 Augt 2015

Edahiro J (2014, January 7) Renewable energy in Japan—current trends show promise and opportunities. Japan For Sustainability (JFS). http://www.ecology.com/2014/01/07/renewable-energy-in-japan-current/. Accessed 31 Aug 2015

Energiewende Germany (2015) Chapter 3(K)—Amendments to the renewable energy sources act in 2014. The Energiewende Story. http://energytransition.de/. Accessed 31 Aug 2015

Environmental Commissioner of Ontario [ECO] (2011, March 22) The true cost of renewable energy and conservation. Environmental Commissioner of Ontario (ECO) Blog. http://www.eco.on.ca/blog/2011/03/22/the-true-cost-of-renewable-energy-and-conservation/. Accessed 31 Aug 2015

Harlan C (2013, June 18) After Fukushima, Japan beginning to see the light in solar energy. The Guardian. http://www.theguardian.com/world/2013/jun/18/japan-solar-energy-fukushima-nuclear-renewable-abe. Accessed 31 Aug 2015

Howlett K, Ladurantaye S (2011, October 7) How McGuinty's green-energy policy cost him a majority. The Globe and Mail. http://www.theglobeandmail.com/news/politics/how-mcguintys-green-energy-policy-cost-him-a-majority-in-ontario/article556454/. Accessed 31 Aug 2015

Independent Electricity System Operator [IESO] (2015) Feed-in tariff program. http://fit.powerauthority.on.ca/what-feed-tariff-program. Accessed 31 Aug 2015

International Energy Agency [IEA] (2014). National survey report of PV power applications in japan—2013. International Energy Agency (IEA) Photovoltaic Power Systems Programme. http://www.iea-pvps.org/index.php?id=146. Accessed 31 Aug 2015

International Energy Agency [IEA] (2015) A snapshot of global markets 2014. Photovoltaic Power Systems Programme. http://www.iea-pvps.org/. Accessed 31 Aug 2015

International Renewable Energy Agency [IRENA] Joint Policies and Measures Database (2015a) Chile—Non-conventional renewable energy law (Law 20.257). http://www.iea.org/policiesandmeasures/renewableenergy/index.php. Accessed 31 Aug 2015

International Renewable Energy Agency [IRENA] Joint Policies and Measures Database (2015b) Canada—Ontario feed-in tariff program. http://www.iea.org/policiesandmeasures/renewableenergy/index.php. Accessed 31 Aug 2015

International Renewable Energy Agency [IRENA] Joint Policies and Measures Database (2015c) United States—State-level renewable portfolio standards. http://www.iea.org/policiesandmeasures/renewableenergy/index.php. Accessed 31 Aug 2015

International Renewable Energy Agency [IRENA] Joint Policies and Measures Database (2015d) Chile—Support for non-conventional renewable energy development program. http://www.iea.org/policiesandmeasures/renewableenergy/index.php. Accessed 31 Aug 2015

International Renewable Energy Agency [IRENA] Joint Policies and Measures Database (2015e) Chile—Non-conventional renewable energy law (Law 20.257). http://www.iea.org/policiesandmeasures/renewableenergy/index.php. Accessed 31 Aug 2015

International Renewable Energy Agency [IRENA] Joint Policies and Measures Database (2015f) Germany—Electricity feed-in law of 1991 ("Stromeinspeisungsgesetz"). http://www.iea.org/policiesandmeasures/renewableenergy/index.php. Accessed 31 Aug 2015

International Renewable Energy Agency [IRENA] Joint Policies and Measures Database (2015g) Germany—Renewable energy sources act (Erneuerbare-Energien-Gesetz (EEG)). http://www.iea.org/policiesandmeasures/renewableenergy/index.php. Accessed 31 Aug 2015

International Renewable Energy Agency [IRENA] Joint Policies and Measures Database (2015h) Germany—2009 amendment of the renewable energy sources act (EEG). http://www.iea.org/policiesandmeasures/renewableenergy/index.php. Accessed 31 Aug 2015

International Renewable Energy Agency [IRENA] Joint Policies and Measures Database (2015i) Australia—Carbon pricing mechanism. http://www.iea.org/policiesandmeasures/renewableenergy/index.php. Accessed 31 Aug 2015

International Renewable Energy Agency [IRENA] Joint Policies and Measures Database (2015j) Indonesia—Old geothermal law (Law No. 27/2003). http://www.iea.org/policiesandmeasures/renewableenergy/index.php. Accessed 31 Aug 2015

International Renewable Energy Agency [IRENA] Joint Policies and Measures Database (2015k) Indonesia—New geothermal law (Law No. 21/2014). http://www.iea.org/policiesandmeasures/renewableenergy/index.php. Accessed 31 Aug 2015

International Renewable Energy Agency [IRENA] Joint Policies and Measures Database (2015l) Japan—Law and establishment of NEDO. http://www.iea.org/policiesandmeasures/renewableenergy/index.php. Accessed 31 Aug 2015

International Renewable Energy Agency [IRENA] Joint Policies and Measures Database (2015m) Japan—New purchase system for solar power-generated electricity. http://www.iea.org/policiesandmeasures/renewableenergy/index.php. Accessed 31 Aug 2015

Islam S (2012) Increasing wind energy penetration level using pumped hydro storage in island micro-grid system. Int J Energy Environ Eng 3:9. doi:https://doi.org/10.1186/2251-6832-3-9. http://www.journal-ijeee.com/content/3/1/9

Jankowska K (2014) Chapter 13—the German policy support mechanism for photovoltaics: the road to grid parity. In: Moe E, Midford P (eds) The political economy of renewable energy and energy security: common challenges and national responses in Japan. Palgrave Macmillan, China and Northern Europe, p 268

Kanargelidis G, Wong S, Libbey A (2013, July 4) Ontario's FIT program: implications of WTO decision. Blake, Cassels & Graydon LLP. http://www.blakes.com/mobile/bulletins/pages/details.aspx?bulletinid=1767. Accessed 31 Aug 2015

KfW Group [KfW] (n.d.) Renewable energy information sheet. https://www.kfw.de/Download-Center/F%C3%B6rderprogramme-(Inlandsf%C3%B6rderung)/PDF-Dokumente/6000002171-M-Offshore-Windenergie-englisch.pdf. Accessed 31 Aug 2015

Lopez BD (2015, January 9) Ecuador: cumulative PV power reaches 26 MW. PV Magazine. http://www.pv-magazine.com/news/details/beitrag/ecuador--cumulative-pv-power-reaches-26-mw_100017984/. Accessed 31 Aug 2015

Meier P, Randle JB, Lawless JV (2015) Unlocking Indonesia's geothermal potential. Joint report of the World Bank (WB) and Asian Development Bank (ADB). http://www.adb.org/sites/default/files/publication/157824/unlocking-indonesias-geothermal-potential.pdf. Accessed 31 Aug 2015

Mendona M (2007) Part 2—policies around the world: Japan. In: Feed-in tariffs: accelerating the deployment of renewable energy. World Future Council. Cromwell Press, Trowbridge, p 73

NaiKun Wind Energy Group Incorporated [NaiKun] (2015) The project—Engineering. http://naikun.ca/the-project/engineering/. Accessed 31 Aug 2015

Nakakuki S, Kudo H (2003) Discussion point in Japan's renewable energy promotion policy: effect, Impact and Issues of the Japanese RPS. Institute of Energy Economics, Japan (IEEJ). 382nd Regular Researchers' Meeting, June 2003 https://eneken.ieej.or.jp/en/data/pdf/205.pdf. Accessed 31 Aug 2015

Nalcor Energy [Nalcor] (n.d.) Nalcor operations—Ramea. http://nalcorenergy.com/ramea.asp. Accessed 31 Aug 2015

Natural Resources Canada [NRCan] (2014) Ramea Island. http://www.nrcan.gc.ca/energy/renewable-electricity/wind/7319. Accessed 31 Aug 2015

Nielsen S (2013, October 14) Chile doubles renewable energy goal to 20% to spark new projects. Bloomberg Business. http://www.bloomberg.com/news/articles/2013-10-14/chile-doubles-renewable-energy-goal-to-20-to-spark-new-projects. Accessed 31 Aug 2015

Non-Conventional Renewable Energy Law (2008) Chilean Law 20.257

Nord Pool Spot (2015) The nordic electricity exchange and the nordic model for a liberalized electricity market. http://nordpoolspot.com/globalassets/download-center/rules-and-regulations/the-nordic-electricity-exchange-and-the-nordic-model-for-a-liberalized-electricity-market.pdf. Accessed 31 Aug 2015

Ontario Energy Board [OEB] Market Surveillance Panel (2012) Monitoring Report on the IESO-Administered Electricity Markets for the period from May 2011–October 2011. Ontario Energy Board (OEB). http://www.ontarioenergyboard.ca/OEB/_Documents/MSP/MSP_Report_20120427.pdf. Accessed 31 Aug 2015

Orkustofnun (n.d.-a) Geothermal. http://www.nea.is/geothermal/. Accessed 31 Aug 2015

Orkustofnun (n.d.-b) Geothermal—Legal and regulatory framework. http://www.nea.is/geothermal/legal-and-regulatory-framework/nr/102. Accessed 31 Aug 2015

PEI Energy Corporation, Island Wind Energy Securing our Future: The 10 Point Plan

Renewable Energy Sources (RES) Act 2014. Statutes of Germany, 2014. Unofficial translation. http://www.bmwi.de/English/Redaktion/Pdf/renewable-energy-sources-act-eeg-2014,property=pdf,bereich=bmwi2012,sprache=en,rwb=true.pdf. Accessed 31 Aug 2015

Saastamoinen M (2009a) Case study 18—Samso: renewable energy island. Changing Behaviour Project. http://www.energychange.info/casestudies/175-samso-renewable-energy-island. Accessed 31 Aug 2015

Saastamoinen M (2009b) Case Study 18—Samso: renewable energy island. Changing Behaviour Project. http://www.energychange.info/casestudies/175-samso-renewable-energy-island. Accessed 31 Aug 2015

Sastrawijaya K, Kurniawan S, Tobing M (2014) New Geothermal Law Finally Issued. Hadiputranto, Hadinoto & Partners. http://www.hhp.co.id/files/Uploads/Documents/Type%202/HHP/al_jakarta_newgeothermallaw_oct14.pdf. Accessed 31 Aug 2015

Short Law I, Chilean Law 19.940

Simms D (2011, March 30) NaiKun headaches hold lessons for offshore wind projects. Canadian Broadcasting Corporation (CBC). http://www.cbc.ca/news/canada/naikun-headaches-hold-lessons-for-offshore-wind-projects-1.992837. Accessed 31 Aug 2015

Starn J, Zha W (2015, April 9) As Germans block Danish wind, a new feud tests crisis-weary EU. Bloomberg Business. http://www.bloomberg.com/news/articles/2015-04-09/as-germans-block-danish-wind-a-new-feud-tests-crisis-weary-eu. Accessed 31 Aug 2015

Thomas R (2000) Chapter 3—legitimate expectations. In: Legitimate expectations and proportionality in administrative law. Hart Publishing, Portland, p 42

Timmins T, Blumer S (2013, June 18) Canada: Ontario's energy minister announces changes to feed-in tariff program. Gowling Lafleur Henderson LLP. http://www.mondaq.com/canada/x/245550/Renewables/Ontarios+Minister+Of+Energy+Announces+Changes+To+FeedIn+Tariff+Program. Accessed 31 Aug 2015

Timmins T, Kirsh M (2012) Ontario's FIT program: domestic content requirements. Gowling Lafleur Henderson LLP. http://www.gowlings.com/knowledgecentre/publicationPDFs/20120615_Gowlings-Ontarios-FIT-Domestic-Content-Requirements_EN.pdf. Accessed 31 Aug 2015

Transparency International (2015) Corruption perceptions index: results. https://www.transparency.org/cpi2014/results. Accessed 31 Aug 2015

Vestas Wind Systems A/S [Vestas] (2015) Company profile. https://www.vestas.com/en/about/profile#. Accessed 31 Aug 2015

Visit Samso (2015) Facts about Samso. http://www.visitsamsoe.dk/en/inspiration/facts-about-samsoe/. Accessed 31 Aug 2015

von Hatzfeldt S (2013) Renewable energy in Chile: barriers and the role of public policy. J Int Aff 66(2):–199. http://jia.sipa.columbia.edu/renewable-energy-chile/. Accessed 31 Aug 2015

Watanabe C (2015, June 17) Fukushima wind project to add largest turbine ever used at sea. Bloomberg Business. http://www.bloomberg.com/news/articles/2015-06-17/fukushima-wind-project-to-add-largest-turbine-ever-used-at-sea. Accessed 31 Aug 2015

World Trade Organization [WTO] (2012) Reports of the panels: 'Canada—Certain measures affecting the renewable energy generation sector' & 'Canada—Measures relating to the feed-in tariff program', joint WTO rulings WT/DS412/R & WT/DS426/R, p 56. World Trade Organization (WTO) Documents Online Database. https://docs.wto.org/dol2fe/Pages/FE_Search/FE_S_S005.aspx. Accessed 31 Aug 2015

Yirka B (2013, January 18) Japan to replace nuclear plant with world's largest wind farm. Phys.org. http://phys.org/news/2013-01-japan-nuclear-world-largest-farm.html. Accessed 31 Aug 2015

Chapter 6
Using Life Cycle Assessment to Facilitate Energy Mix Planning in the Galapagos Islands

Eduard Cubi, Joule Bergerson, and Anil Mehrotra

Introduction

The population in the Galapagos Islands has increased by one order of magnitude in the last 50 years (from 2400 in 1962 to 26,600 in 2012) largely due to demographics and a relatively strong economy based on tourism. In turn, the annual number of tourists has increased from 20,000 in 1980 to 140,000 in 2006 (Carrion 2007). These increases translate into a sharp increase in energy use in the archipelago. Total primary energy consumed in the Galapagos in 2006 was approximately 1.5×10^6 GJ, which is 1.5% of the annual energy use in Ecuador. Eighty percent of the energy consumed in the Galapagos was in the form of fuel for transportation (diesel and gasoline used in ships and motor boats), while the remaining 20% was diesel use for electricity generation. Diesel and gasoline are shipped by tankers from mainland Ecuador through very sensitive marine ecosystem surrounding the Islands. Historically, there have been several oil spills in the Galapagos, with the most significant being the spill from the Tanker "Jessica" in 2001.

This released 175,000 gallons of diesel and killed 60% of the iguanas in the archipelago (Hecht 2002; Lewis 2013; Lewis and Galapaface 2014). In addition to a high risk of oil spills and subsequent environmental and economic costs, using a GHG-intensive energy carrier such as diesel for both marine transportation and electricity generation contributes to climate change. The Galapagos is likely to be affected by sea level rise in the context of climate change. This, combined with its

---

E. Cubi • J. Bergerson (✉) • A. Mehrotra
Energy and Environment Systems Group (EESG), Centre for Environmental Engineering Research and Education (CEERE), Department of Chemical and Petroleum Engineering, Schulich School of Engineering, University of Calgary, Calgary, AB, Canada
e-mail: jbergers@ucalgary.ca

© Springer International Publishing AG 2018
M.-E. Tyler (ed.), *Sustainable Energy Mix in Fragile Environments*,
Social and Ecological Interactions in the Galapagos Islands,
https://doi.org/10.1007/978-3-319-69399-6_6

high influx of international visitors, makes the Galapagos a good candidate to become a world leader in responsible decision-making related to energy systems.

Since 2004, initiatives have been in place to reduce diesel import and use in the Galapagos. Over the past 15 years, the ERGAL ("Energias Renovables para las Galapagos", Renewable Energy for the Galapagos) project has been a collaborative initiative (Ministerio de Electricidad y Energia Renovable et al. 2015) of the Ministerio de Electricidad y Energia Renovable, the United Nations Development Program, the Global Environmental Facility and the Consejo Nacional de Electricidad. ERGAL promotes renewable electricity generation projects in the archipelago, but to date has not addressed energy use in marine transportation which is the main source of diesel imports. In order to enable more informed energy policy decisions that account for the full environmental and cost implications of energy supply and use in the Galapagos, alternatives to diesel and gasoline imports should be identified and assessed with a consistent set of criteria and include a wide range of environmental and economic impacts. Life cycle assessment (LCA) is a method grounded in decision analysis that can be used to evaluate the environmental impacts of a product or process from the initial extraction of resources through to the disposal of unwanted residuals.

The objective of this paper is to introduce LCA as a relevant tool to inform energy system decision-makers about sustainable energy mix potential. Energy mix is defined as a combination of energy sources and conversions that create a mix of energy supply pathways to meet the demand for energy services in a given location, which, in this case, is the Galapagos Islands. The potential value of LCA is illustrated using a high-level assessment of the GHG emissions associated with several energy options in the Galapagos Islands which provides first approximations of costs, a qualitative discussion marine spill risks and recommendations for future LCA use.

An Introductory Summary of Life Cycle Assessment

Life cycle assessment (LCA) is "a systematic set of procedures for compiling and examining the inputs and outputs of materials and energy and the associated environmental impacts directly attributable to the functioning of a product or service system throughout its life cycle" (International Organization for Standardization 2010). The three main stages in a life cycle assessment are:

- Goal and Scope Definition
- Life Cycle Inventory (LCI)
- Impact Assessment

Goal and Scope Definition

This is an explicit statement of the purpose and scope of the study. The scope describes the specific characteristics of the study, which must be defined consistent with the intended application. For example, the assessment method must be defined so that the results can respond to the research question. Technical details identified in the scope definition should include:

- The functional unit which quantifies the service delivered by a product or a service and is the basis of comparison for the different options under assessment
- The system boundaries such as life cycle stages and processes included in the assessment
- The impact categories (types of environmental impacts) considered
- The level of detail, assumptions, and limitations

Life Cycle Inventory (LCI)

This is the quantification of the energy, raw material, and water inputs as well as the outputs to air, land and water associated with each life cycle stage of a product or a service. The LCI usually involves the development of a model of the technical system under study (including energy and mass balance for the system) and extensive data collection and calculations. While the model and derived calculations are incomplete simplifications of the actual system, they should include all the environmentally relevant flows identified in the initial LCA goal and scope definition.

Impact Assessment

Translates the LCI input and output flow results such as litres or gallons of diesel spilled into potential environmental impacts (such as toxicity to aquatic life).

As such, LCA results can be used to compare alternative products, technologies or pathways for the same service (functional unit) and identify opportunities to reduce energy and material inputs or environmental impacts at the different stages of product or service life cycle.

Goal and Scope Identification for Energy Supply Options to the Galapagos

The objective is to compare the environmental impacts and economic viability of a set of options to replace diesel shipments into the Galapagos Islands. This analysis is meant to facilitate further dialogue within the region about the trade-offs involved in sustainable energy system decisions. The scope includes total annual fuel shipments into the Galapagos for both power generation and transportation. Specifically, the functional unit is 1 year of total energy supply in the Galapagos Islands (1.5×10^6 GJ/yr) (Carrion 2007).

The environmental impacts and costs of different energy mix alternatives are assessed from "cradle to grave" or from extraction of resources through to end use. The GHG emissions for the alternatives energy pathways identified (natural gas, gas to liquids and Jatropha biodiesel) were assessed quantitatively. The costs are discussed but have a high level of uncertainty. Potential risks associated with marine spills are also discussed qualitatively.

The Energy Mix Options

For LCA demonstration purposes, current diesel and gasoline imports for electricity generation in the Galapagos are compared with two alternative energy pathways:

- Natural gas imports for electricity generation in the Galapagos and gas to liquids conversion in the Galapagos for transportation purposes
- Biodiesel imports from mainland Ecuador for electricity generation and transportation purposes in the Galapagos

In order to compare these two alternatives with the current energy baseline, the assessment of environmental impacts and costs of these three pathways (the two alternatives + baseline) include the same life cycle stages from extraction of resources to energy end use in the Galapagos.

The specific details of the analysis, such as all the numeric assumptions associated with each pathway, are not provided in this paper. Instead, the pathways are described qualitatively, starting with some background on resource availability in the Galapagos and Ecuador and following with the assumed energy transformations up to the final energy use in the Galapagos Islands. Only the main sources of data are cited. Each of the three pathways identified together with the main assumptions made in LCA assessment are described as follows.

Pathway One: Baseline Diesel and Gasoline Imports from Ecuador Mainland

Ecuador is a net exporter of crude oil. However, due to its very limited refinery capacity, it is a net importer of refined oil products, mainly from the United States (Energy Information Administration 2014). The baseline pathway assumes that crude oil is extracted in Ecuador and fuel products (diesel and gasoline) are produced in refineries located in the United States, shipped to mainland Ecuador and then to the Galapagos Islands by marine tanker. Diesel is then combusted in the Galapagos for electricity generation and in internal combustion engines for transportation fuel. Gasoline is consumed in the Galapagos for transportation fuel only by small boats.

Total energy demand (diesel and gasoline) in the Galapagos is based on 2006 data (Carrion 2007). Estimates of upstream emissions for diesel and gasoline production and emissions from freight transportation of liquid fuels are based on Natural Resource Canada GHG calculations (Natural Resources Canada 2014). Emission factors for diesel and gasoline using GHG emissions per unit of energy are based on US Environmental Protection Agency guidelines (US Environmental Protection Agency 2004).

Pathway Two: Natural Gas and Gas to Liquids

This pathway assumes that natural gas (as opposed to diesel) is the main energy source is shipped into the Galapagos Islands. Compared to other fuels, natural gas is relatively cheap and clean with the lowest GHG emissions per unit of energy among fossil fuels (US Environmental Protection Agency 2004). Liquefied natural gas (LNG) markets have been growing quickly and making natural gas a real option even in relatively remote areas.

Ecuador has relatively small proven natural gas reserves and a very limited natural gas market (Energy Information Administration 2014). Low natural gas utilization rates in Ecuador are due mainly to a lack of infrastructure to capture and market natural gas. Therefore, in a natural gas-based scenario, the original source of natural gas would likely be a natural gas exporter such as the United States or Canada. There are different alternatives for transporting natural gas. Liquefied natural gas (LNG) tankers are most common in marine natural gas transportation. LNG requires capital intensive liquefaction and regasification facilities, but the very high volume ratio of liquefied natural gas (approximately 610:1) makes the transportation by tanker itself very effective in terms of cost per MJ of natural gas transported per kilometre. LNG is a preferred option for large-scale long distance natural gas transportation projects due to the high fixed costs relative to variable costs in LNG transportation systems (Energy Information Administration 2003; US Department of Energy 2005). In contrast, compressed natural gas (CNG) systems have a much lower volume ratio (as low as 2:1). CNG does not require expensive infrastructure at both ends of the supply chain (liquefaction and regasification), but the low volume ratio makes vessel transportation less effective. Therefore, CNG systems are generally suitable for small scale regional gas delivery projects (EnerSea Transport LLC 2015; SeaNG—Coselle 2015).

Even if natural gas was used to replace the total diesel and gasoline consumption, the annual demand of natural gas in the Galapagos would be smaller than the capacity of a single (typical) LNG tanker. Therefore, this pathway assumes that natural gas replaces oil for electricity generation on the Ecuador mainland as well as in the Galapagos. In order to make the natural gas option more feasible, this assumption increases natural gas demand by approximately a factor of 70 compared to the demand in the Galapagos alone. This pathway assumes that natural gas is imported from Canada to mainland Ecuador in the form of LNG. GHG emissions and costs of LNG transportation are allocated based on the natural gas demands in the Galapagos and mainland Ecuador, respectively. This pathway also assumes natural gas will be transported from mainland Ecuador to the Galapagos Islands by compressed natural gas (CNG) vessels. A fraction of the natural gas used in the Galapagos is converted to liquid fuels for transportation purposes, and the remainder is used for electricity generation by natural gas-fired power plants.

Total energy demand using natural gas as a substitute for diesel and gasoline (Carrion 2007). Similarly, estimates for upstream emissions for natural gas are based on Natural Resource Canada (Natural Resources Canada 2014), and emissions from natural gas liquefaction, transportation as LNG and regasification are based on a variety of sources (Jaramillo et al. 2007; Hasan et al. 2009; Abrahams et al. 2015). Emissions associated with the gas-to-liquids conversion and final use are based on Forman et al. (2011). Emissions from natural gas combustion for electricity generation and transportation are based on US Environmental Protection Agency calculations (US Environmental Protection Agency 2004).

Pathway Three: Jatropha Biodiesel

The third alternative pathway evaluates the potential use of biodiesel to replace diesel as the main energy carrier into the Galapagos. Biofuels absorb CO_2 during feedstock growth which reduces their life cycle GHG emissions. *Jatropha curcas* ("pinon") is a bush that grows naturally and abundantly in mainland Ecuador and particularly in the relatively poor area of Manabi. One of the projects within the ERGAL initiative aims to retrofit diesel engines for electricity production so that they can run on pure Jatropha oil (Deutscher Entwicklungsdienst and Vereinigte Werkstätten für Pflanzenöltechnologie 2008). However, extending this initiative to retrofit diesel engines in ships and boats may not be operationally feasible as these ships are not publicly owned. It would be technically challenging to retrofit internal combustion engines. Therefore, this pathway assumes the use Jatropha biodiesel instead of pure Jatropha oil and Jatropha biodiesel which would be shipped from the mainland to the Galapagos Islands for transportation fuel and electricity generation. Total energy demand of biodiesel as a substitute for diesel and gasoline in the Galapagos and estimates of emissions associated with Jatropha production and biodiesel production are based on Carrion (2007) and Whitaker and Heath (2009). Emissions from marine fuel transportation and emission factors from biodiesel combustion have been taken from Natural Resources Canada (2014) and US Environmental Protection Agency (2004). Figure 6.1 summarizes the assumed steps and energy transformations in the three energy supply pathways assessed using LCA.

6 Using Life Cycle Assessment to Facilitate Energy Mix Planning in the Galapagos…

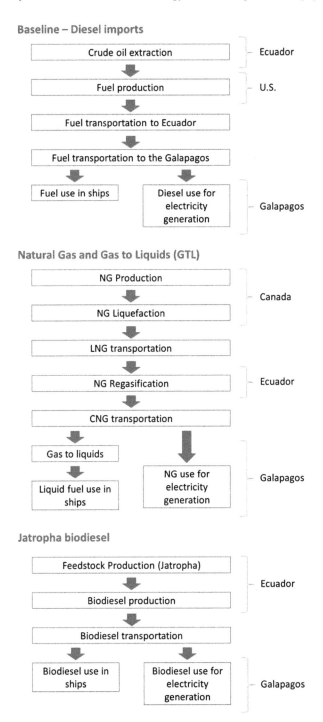

Fig. 6.1 Three pathways for LCA demonstration

Life Cycle Inventory Sample Calculation

A model for each of the energy mix options was developed in the life cycle inventory (LCI) stage of the LCA. For the purposes of illustrating the potential for an LCA approach to sustainable energy mix design, the general method used for LCI calculation of the CO_2 emissions from electricity generation in pathway one (baseline) is described as follows. The baseline pathway (pathway one) assumes that diesel is burned in distributed electricity generation units across the Galapagos Islands, and the quantification of the associated CO_2 emissions can be described as follows:

- Annual diesel use (2006 data) for electricity generation in the Galapagos is obtained from (Carrion 2007):
  - 2.7×10^{11} BTU_{diesel}/yr
- Diesel combustion GHG emissions factor is obtained from (US Environmental Protection Agency 2004):
  - 22.23 lb CO_2/$Gallon_{diesel}$
- CO_2 emissions from electricity generation are calculated combining the factors above and unit conversions (where d is diesel):

CO_2 Emissions from electricity generation

$$= \frac{2.7 \times 10^{11} BTH_d}{yr} \times \frac{l_d}{37.3\ MJ_d} \times \frac{MJ}{947 BTU} \times \frac{22.23\ lb\ CO_2}{Gallon_d} \times \frac{0.454\ kg\ CO_2}{lb\ CO_2} \times \frac{3.785\ l}{Gallon}$$

$$= 2.04 \times 10^7 \frac{kg\ CO_2}{yr}$$

In this example, the contribution of diesel combustion for electricity generation in the total life cycle GHG emissions of this pathway is 2.04×10^7 kg CO_2/yr which is shown in the stacked columns in Fig. 6.2 ("Baseline diesel"/"Use for electricity generation").

LCA GHG Results

Figure 6.2 compares the life cycle GHG emissions associated with the three pathways evaluated. Figure 6.2 also shows the cumulative GHG emissions of each pathway as well as the contributions of the individual phases.

Emissions of the three pathways are dominated by fuel use for transportation (61%) and electricity generation (15%). The upstream emissions (24%) are largely in the crude extraction and diesel production stages, while fuel transportation has a minimal contribution to overall GHG emissions.

6 Using Life Cycle Assessment to Facilitate Energy Mix Planning in the Galapagos...   101

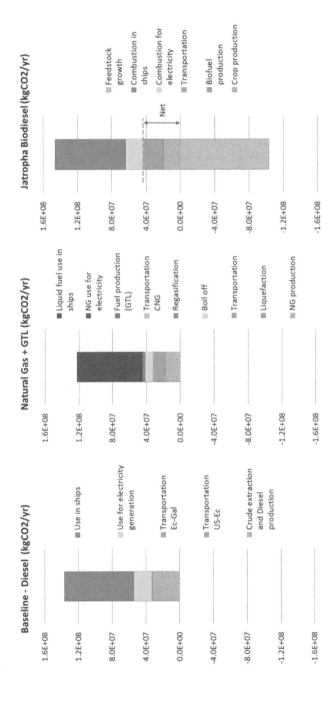

Fig. 6.2 Life cycle GHG emissions

Relative to the baseline pathway, the natural gas pathway shows lower GHG emissions in the use phase (electricity generation and use for transportation) in both relative and absolute terms. Combustion of natural gas appears to be less GHG intensive than the combustion of diesel or gasoline. In contrast, the complex supply chain results in higher emissions in the upstream stages, which add up to 37% of the life cycle GHG emissions. Among the upstream stages, natural gas liquefaction and natural gas, which boil off during LNG transportation, are the largest sources of GHG emissions. All stages considered, the life cycle GHG emissions of the natural gas pathway are approximately 10% lower than the baseline pathway.

In absolute terms, the Jatropha biodiesel pathway has the same emissions in the use phase (fuel combustion for electricity generation and transportation) as the baseline pathway. Upstream GHG emissions from Jatropha crop production and biofuel production add up to 14% and 16% of the total GHG emissions, respectively. This is higher than the upstream emissions in the baseline pathway, but the CO_2 absorbed during feedstock growth offsets approximately 71% of GHG emissions. Therefore, the net life cycle GHG emissions of the Jatropha biodiesel pathway are approximately 69% lower than the baseline pathway.

Discussion

The results of this high-level assessment of three alternative energy pathways for the Galapagos Islands suggest that Jatropha biodiesel has the lowest GHG emissions (69% lower than the baseline), while the natural gas pathway results in moderate GHG reductions relative to baseline (10% GHG emissions reduction). However, to fully understand energy mix choice implications, all three pathways will require a more comprehensive evaluation to include costs, performance and other social and environmental implications that go beyond the GHG emission scenarios presented in this paper.

Preliminary cost calculations suggest that natural gas pathway costs are approximately three times higher than baseline. This can be attributed to the high costs of liquefied natural gas (LNG) and compressed natural gas (CNG) transportation infrastructure for the relatively small energy demand in the Galapagos (Economides 2005). Should these preliminary cost estimates be confirmed by further inquiry, then the natural gas pathway would likely prove to be prohibitively expensive. In view of the potential environmental benefits and high economic costs of natural gas imports into the Galapagos, natural gas production in Ecuador could become a relevant alternative. Domestic natural gas production in Ecuador would require building new infrastructure, but this pathway would save the costs and GHG emissions associated with LNG transportation.

Costs of the Jatropha biodiesel pathway could be similar to or possibly lower than the baseline pathway. Feedstock production accounts for approximately 75% of the biodiesel cost and is the driver of biodiesel economics (Koh and Mohd. Ghazi 2011; Leduc et al. 2009; Cynthia and Teong 2011). Land and water requirements are also need to be calculated into the Jatropha biodiesel pathway. For example, a preliminary estimate suggests that the land use requirements for crop production to satisfy both the annual electricity and the transportation fuel demands could be in the 20–150 km^2 range. This large range is due to the huge discrepancies in information sources which are hard to rationalize because of unstated assumptions and conditions related to these multiple source figures. Additional concerns include potential distortions in the bio markets related to farmers replacing food crops for fuel crops which could contribute to a decrease in local food supply and an increase in costs.

Risks of marine spills during fuel transportation exist for all three pathways. Natural gas spills would pose relatively low risks to the marine ecosystem compared to the diesel pathway (Sandia National Laboratories 2004). Biodiesel spills would generally be less toxic to marine plants and animals and have a faster biodegradation rate than diesel. However, as with diesel spills, birds, mammals and fish that get coated with biodiesel are at risk from hypothermia, food source contamination and vulnerability to predators (Von Wedel 1999).

Conclusions and Future Work

Life cycle assessment is a potential tool for decision-makers in designing sustainable energy mix solutions. As illustrated in this paper, LCA has the potential to identify advantages and disadvantages of alternative energy source pathways. However, it requires a much more rigorous analysis than what is possible with currently available data. In order to fully utilize LCA in energy mix planning, the following information requirements need to be addressed through further work:

- A more comprehensive understanding of the local Galapagos Islands marine and terrestrial environmental concerns and the risks associated with natural gas and biodiesel spills is required. Biodiesel or natural gas would in theory reduce risks associated with diesel marine spills. The relative benefits could not be quantified, and it is not clear if natural gas or biodiesel spills might have other impacts.
- The social ecological and economic factors associated with the land and water requirements of growing Jatropha in mainland Ecuador need to be identified in collaboration with key local informants. The potential land and water use impacts associated with growing Jatropha could not be assessed without this information.
- Future research needs to identify the cost factors and costs affecting the economic competitiveness of different energy sources and opportunities to improve cost effectiveness in terms of \$/kg CO_2. The cost estimates used for this LCA demonstration are highly uncertain and need to be more fully understood.

Further collaboration with government, public and private sector stakeholders in both the Galapagos Islands and Ecuador is an important next step in identifying relevant energy pathways and sustainable energy mix solutions as well as assembling the data and information necessary to enable a more detailed and comprehensive LCA analysis. However, in spite of the limitations in this initial exploration, life cycle assessment illustrated some important insights into sustainable energy mix planning:

- Energy mix options can be assessed from cradle to grave and with a consistent set of parameters and enable a comparative and clear comparison across alternatives.
- Energy pathways with the highest potential for GHG emissions reductions can be identified in a relative context and compared to current baseline conditions.
- The identification of life cycle stages with the highest impacts and costs can assist in identifying opportunities for performance improvement in selected pathways.

The LCA approach as illustrated in this paper is a valuable tool that can inform energy policy makers in evaluating sustainable energy mix options and offers an approach to better understanding both the energy needs and the possible energy mix solutions for the Galapagos.

References

Abrahams LS, Samaras C, Griffin WM, Matthews HS (2015) Life cycle greenhouse gas emissions from U.S. liquefied natural gas exports: implications for end uses. Environ Sci Technol 49:3237–3245

Carrion GC (2007) Estudio de prevision de la demanda de energia para las Islas Galapagos: escenarios socioeconomicos. Available from: http://www.ergal.org/cms.php?c=1276

Cynthia O-B, Teong LK (2011) Feasibility of Jatropha oil for biodiesel: economic analysis. In: World renewable energy congress, Linkoping

Deutscher Entwicklungsdienst, Vereinigte Werkstätten für Pflanzenöltechnologie (2008) Energia Renovable para Galapagos. Sustitucion de combustibles fosiles por biocombustibles en la generacion de energia electrica en la Isla Floreana. Available from: http://www.ergal.org/cms.php?c=1272

Economides MJ (2005) The economics of gas to liquids compared to liquefied natural gas. World Energy 8:136–140

Energy Information Administration (2003) The global liquefied natural gas market: status and outlook. Available from: http://www.eia.gov/oiaf/analysispaper/global/pdf/eia_0637.pdf

Energy Information Administration (2014) Ecuador—overview. Available from: http://www.eia.gov/countries/cab.cfm?fips=ec

EnerSea Transport LLC (2015) Understanding CNG [Internet]. Available from: http://enersea.com/understanding-cng/

Forman GS, Hahn TE, Jensen SD (2011) Greenhouse gas emission evaluation of the GTL pathway. Environ Sci Technol 45:9084–9092

Hasan MMF, Zheng AM, Karimi IA (2009) Minimizing boil-off losses in liquefied natural gas transportation. Ind Eng Chem Res 48:9571–9580

Hecht J (2002) Galapagos oil spill devastated marine iguanas. New Scientist

International Organization for Standardization (2010) ISO 14042:2000 Environmental management—Life Cycle Assessment—Life Cycle Impact Assessment

Jaramillo P, Griffin WM, Matthews HS (2007) Comparative life-cycle air emissions of coal, domestic natural gas, LNG, and SNG for electricity generation. Environ Sci Technol 41:6290–6296

Koh MY, Mohd. Ghazi TI (2011) A review of biodiesel production from Jatropha curcas L. oil. Renew Sust Energ Rev 15:2240–2251

Leduc S, Natarajan K, Dotzauer E, McCallum I, Obersteiner M (2009) Optimizing biodiesel production in India. Appl Energy 86(Suppl 1):S125–SS31

Lewis G (2013) The struggle for sustainable energy on Galapagos. Galapagos Digital

Lewis G, Galapaface AC (2014) I: challenges in salvage effort. Galapagos Digital

Ministerio de Electricidad y Energia Renovable, United Nations Development Program, Global Environmental Facility, Consejo Nacional de Electricidad (2015) ERGAL—Energias Renovables para Galapagos [Internet]. Available from: http://www.ergal.org/cms.php?c=1233

Natural Resources Canada (2014) GHGenius. A model for lifecycle assessment of transportation fuels

Sandia National Laboratories (2004) Guidance on risk analysis and safety implications of a large liquefied natural gas (LNG) spill over water. Available from: http://www.energy.ca.gov/lng/documents/2004-12_SANDIA-DOE_RISK_ANALYSIS.PDF

SeaNG—Coselle (2015) Compressed natural gas—demand customers [Internet]. Available from: http://www.coselle.com/applications/demand-customers

US Department of Energy (2005) Liquefied natural gas: understanding the basic facts. Available from: http://energy.gov/sites/prod/files/2013/04/f0/LNG_primerupd.pdf

US Environmental Protection Agency (2004) Unit conversions, emissions factors, and other reference data. Available from: http://www.epa.gov/appdstar/pdf/brochure.pdf

Von Wedel R (1999) Technical handbook for marine biodiesel. In Recreational boats, 2nd edition, Cytoculture international, Point Richmond, CA

Whitaker M, Heath G (2009) Life cycle assessment of the use of Jatropha biodiesel in Indian locomotives. Available from: http://www.nrel.gov/biomass/pdfs/44428.pdf

Chapter 7
Sustainability of Renewable Energy Projects in the Amazonian Region

Juan Leonardo Espinoza, José Jara-Alvear, and Luis Urdiales Flores

Introduction

The Amazon region is shared by nine South American countries (Brazil, Peru, Bolivia, Colombia, Ecuador, Guyana, Venezuela, French Guyana, and Suriname). It covers an area of approximately 6,000,000 km^2. The Amazon region is a fragile ecosystem with high biodiversity, and it has a major contribution in mitigating climate change. That is why the region has received the title of "lungs of the planet." Historically, the Amazon region has been inhabited by a great diversity of aboriginal people who have managed to live in harmony with the forest for centuries. However, because of the colonization of the late nineteenth and early twentieth centuries, which has accelerated in recent years with timber extraction, oil, mining, agriculture, livestock, and tourism, the region has been losing steadily native forest, emerging urban centers, and increasing the nonindigenous population. Urbanization in various parts of the region is pressing to provide basic services to the population such as drinking water, sewerage, and electricity.

In the case of Ecuador, there are six Amazonian provinces, Orellana, Pastaza, Napo, Sucumbíos, Morona Santiago, and Zamora Chinchipe, representing an area of 120,000 km^2 (48% of the country size). According to the results of the 2010 Census of Population and Housing, there is an estimated total population of

---

J.L. Espinoza (✉)
Department of Electrical and Electronics-DEET, Faculty of Engineering,
University of Cuenca, Cuenca, Ecuador
e-mail: juan.espinoza@ucuenca.edu.ec

J. Jara-Alvear
Center for Development Research (ZEF), University of Bonn, Bonn, Germany
e-mail: jose.jara.a@gmail.com

L. Urdiales Flores
Empresa Eléctrica Regional Centrosur C.A., Cuenca, Ecuador
e-mail: lurdiales@centrosur.com.ec

© Springer International Publishing AG 2018
M.-E. Tyler (ed.), *Sustainable Energy Mix in Fragile Environments*,
Social and Ecological Interactions in the Galapagos Islands,
https://doi.org/10.1007/978-3-319-69399-6_7

750,000 in the Ecuadorian Amazonian region, which represents 5% of the country total population (INEC 2010). Ten different ethnic groups constitute the rural population in this region (CODENPE 2012). Services coverage at provincial level reaches 55.6% of drinking water and 41.4% of sewerage, well below the national average, which is 72% and 53.6%, respectively (INEC 2010).

In year 2012, the electricity coverage in Ecuador was 96.9% at national level and reached 89.8% of rural population mainly through investment on grid extension (ARCONEL 2013). The government fund program FERUM (Fund for Rural and Urban-Marginal Electrification) financed this investment. Nevertheless, this approach has left behind the most isolated and disadvantage rural communities where a grid extension is unfeasible due to their limited access, grade of dispersion, and low demand. Most of this population is scattered along the Ecuadorian Amazon region. For instance, the electricity coverage in this region reached 88.6% in 2012 (ARCONEL 2013), but if this indicator is disaggregated into urban and rural, the rural electricity coverage was just 72%, almost 18 points below the national average.

Since year 2000, some efforts to solve this problem focused on installing photovoltaic solar home systems (SHS) on Amazonian households, through government and international donor's initiatives (Vasconez 2010). After their implementation, however, there has not been a systematic evaluation making it difficult to know the current technical situation of approximately 3000 systems as well as their real impact on local conditions. This lack of information and the abandonment of many of these projects have hindered the scaling up of decentralized rural electrification (DRE) initiatives in Ecuador where the electric grid is unfeasible.

In 2008, a new constitution took effect establishing the "good living" condition as the main objective for the country. Rural electrification was a national priority in order to contribute to improve the living conditions of rural population. The good living or "sumak kawsay," its translation from the Quechua language, is a indigenous view of the world that focuses on the human being and seeks to meet the needs in order to get a good quality of life. This view proposes to live in peace and harmony with nature and looks for the indefinite prolongation of cultures (SENPLADES 2013). It is clear that the good living has important similarities to the accepted concept of sustainable development as defined by the World Commission on Environment and Development – WCED in 1987.

In both paradigms, good living and sustainable development, energy access has been recognized as an important goal. In that sense, the use of renewable energy in isolated and fragile ecosystem like the Ecuadorian Amazon region is a feasible solution that contributes on improving people's life and reducing negative impacts on the environment.

In 2010, one of the several public electric distribution companies EERCS C.A., known as Centrosur, started the project "Yantsa Ii Etsari" (that translates as "light from our sun" in Shuar language) to electrify, with SHS, 3000 isolated indigenous households (Shuar and Achuar) scattered along the province Morona Santiago in the southern part of the Ecuadorian Amazonian region. After 6 years of continuous operation, 3266 SHS have been installed, covering almost 100% of identified isolated indigenous population. Centrosur operates and maintains the SHS, which demands important human, economic, and technological resources.

This chapter aims to analyze sustainability challenges and prospects for renewable energy projects for electricity service provision in the Amazon region context using the Yantsa Ii Etsari project as a case study. This experience could be relevant for other decision-makers or researchers interested in sustainable energy development in isolated and fragile regions such as the Amazon. The chapter reviews the research projects implemented by Centrosur and academia in order to enhance sustainability of the Yantsa Ii Etsari project in areas like "mobility," "operations and maintenance," "environmental impact assessment," and "policies and procedures." Finally, a discussion on the way forward for sustainability assessment protocols for renewable energy projects is presented.

The Yantsa Ii Etsari Project

Based on literature review of field reports and unstructured interviews to key administrative and technical staff, Jara-Alvear and Urdiales (2014) developed a qualitative analysis of the project Yantsa Ii Etsari from formulation to implementation (Fig. 7.1). The trigger that raised the political will of the Ministry of Electricity and Renewable Energy (MEER) to invest on the project was a human rabies epidemic – spread by bats that affected the study area in 2008. It was believed that electricity access (lighting) will improve local conditions among households, which has also proved a reduction on vector diseases (Mendes et al. 2009). Once national funds were released, 120 isolated communities were selected in an agreement between Centrosur, the National Council of Electricity (CONELEC) (nowdays ARCONEL), and the indigenous organization (FISCH). It was a top-down approach, where the final users participated during the project socialization stage, through workshops, and later on in the installation process.

The electricity needed in a typical indigenous family of the study area was estimated in 322 watts-hour per day (Wh/day), which is supplied by a standardized SHS (Table 7.1). It might appear potential productive uses of electricity for homeowners or communal services that would require more energy. However, one of the advantages of solar photovoltaic systems is their modularity so that additional energy needs can be met with more panels and batteries in the same installation. The standardization facilitated the bidding process and logistics to accomplish the project aims (Jara-Alvear and Urdiales 2014).

Between 2010 and 2015, about 20 local or regional contractors did the transportation and installation of 3266 SHS in almost 200 communities (Table 7.2). This aspect enhanced local professional capabilities. Based on commissioning reports, the average cost of an SHS was $16 per peak watt (Wp) where transportation and installation costs have an important 41.6% share (Jara-Alvear and Urdiales 2014). Table 7.3 shows the average costs of the SHS installation. Once the installation was finished, Centrosur and the household head sign a service contract, where the user

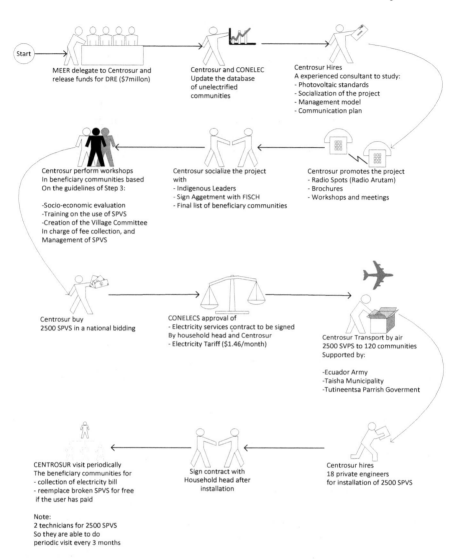

Fig. 7.1 Flowchart of Yantsa Ii Etsari project. Source: Jara-alvear and Urdiales (2014)

has to pay a fixed fee of US$1.46/month and the utility will provide maintenance and continuous services. The utility has to work in coordination with a community electrification committee that provides routine minor maintenance services. Although the fee does not cover the investment, it does raise awareness of aboriginal people on the duty to pay for services. Therefore, the electric utility uses resources from its budget to subsidize transportation, spare parts, and other costs in order to guarantee the financial sustainability of the project.

Table 7.1 Characteristics of SHS

Description	Value
Solar photovoltaic panels	150 Wp
Battery	150 Ah
Solar regulator	20 A
Inverter (12 V/120 V)	300 W
Average solar radiation	4 kWh/m^2/day
Expected production of SHS	400 Wh/day
Load (lights, radio, TV/DVD, battery charger)	322 Wh/day

Source: Jara-Alvear and Urdiales (2014)

Table 7.2 Number of SHS installed by the project

Year	Number of communities	Number of SHS
2011	15	290
2012	108	2063
2013	7	109
2014	34	432
2015	32	372
Total	196	3266

Source: Urdiales (2015)

Table 7.3 Average costs of SHS installation

Description	Cost	Observation
Equipment	$1400	Complete SHS
Transport	$600	Variable cost: Air freights, fuel. Not included community labor
Labor	$400	Variable cost: Labor of installation, workshops and informative campaign
Total	$2400	

Source: Jara-Alvear and Urdiales (2014)

Sustainability Overview of Yantsa Ii Etsari

After the Brundtland report was released in 1987, sustainability has gained attention in academia, industry, and government. The report's definition of sustainability refers to "development that meets present needs without compromising the ability of future generations to meet their own" (WCED 1997). This definition stresses the multidimensional character of sustainability and the equitable distribution of resources. In this regard, Elkington (1999) proposed a "triple bottom line" model, and Mebratu (1998) proposed "the cosmic interdependence" model; both models define sustainability using economic, environmental, and social aspects. These three pillars are widely used on sustainability assessment and policymaking.

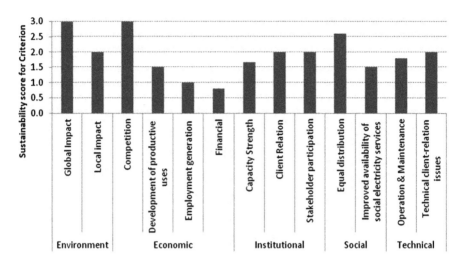

Fig. 7.2 Sustainability dimensions of Yantsa Ii Etsari project

Regarding the energy sector, "sustainable energy development will require electricity services that are reliable, available and affordable for all, on a sustainable basis, world-wide" (Johansson and Goldemberg 2002). Again, this definition implies the consideration of economic, environmental, and social dimensions for sustainability assessment. However, literature (Ilskog 2008; Brent and Rogers 2010) suggests that two additional dimensions should be integrated in rural electrification projects: technical and institutional pillars. These are important since reliability, human capacity, and local management of energy systems are key aspects in rural areas where technology and specifically renewable energy market are in the early stages (Mainali and Silveira 2015).

As part of a doctoral research project in progress (ZEF-University of Bonn and Centrosur), Jara-Alvear and Urdiales (2014) attempted to do a rapid sustainability assessment of Yantsa Ii Etsari project using the framework proposed by Ilskog (2008), which uses 5 dimensions and 13 criteria associated with 37 indicators to capture the complexity of sustainability in rural electrification projects (Fig. 7.2). This framework provided a holistic starting point to discuss project sustainability and guide the research agenda.

Even though Yantsa Ii Etsari represents an alternative to increase Centrosur rural coverage objectives, the project also faces several important challenges. These deal with economic, environmental, institutional, technical, and social dimensions all related to the project sustainability.

Table 7.4 presents a summary of how previous results (Fig. 7.2) have guided the definition of research projects in order to enhance information and knowledge to move toward sustainability in rural electrification in the Ecuadorian Amazon region.

Table 7.4 Sustainability assessment and research projects for Yantsa Ii Etsari

Dimension	Focus for sustainability assessment of decentralized rural electrification projects	Research project
Environmental • Global impact • Local impact	Replacement of polluting source of energy (kerex, candles, dry batteries, diesel, gasoline). Waste management system during installation, operation, and decommission At global scale, reducing the contribution of greenhouse emissions and the indirect impact on Amazon forest conservation	Integrated waste management system
Economic • Competition • Development of productive uses • Employment generation • Financial	High cost of installation and maintenance, and the fixed fee ($1.46/month) established by existing regulation, makes the project not profitable. Financial mechanism, efficient subsidy mechanism, and reduction of maintenance cost are key issues to make rural electrification with SHS less dependent on external source of funding or subsidies. Accessibility and transportation costs (air, river) have an important influence on the final cost of energy production but also on socioeconomic development of indigenous people	Development of a solar boat prototype Design of an automatic reliability centered Maintenance model
Institutional • Capacity strength • Client relation • Stakeholder participation	Capacity strength of the electric company and stakeholder participation are interconnected issues that need to be addressed. There is a need to strengthening the existing multicultural stakeholder network made of electricity company staff, local technicians, authorities, and clients Being modern energy a new actor in daily activities of people, "the rules of the game" can change dramatically in communities. This new institutional order is indeed a challenge that all project stakeholders are facing	Local utility involvement and stakeholder participation
Technical • Operation and maintenance • Technical client-relation issues	Operation and maintenance need to improve. The failure rate of SHS jeopardizes not only project economy but also the environment due to electronic waste generation. Increasing the reliability of SHS and adopting an integrated asset management strategy could help to reduce maintenance cost and enhance the service quality. However, during the design phase and planning, quality of studies and reliability of information are crucial to ensure suitable solutions for local context	Reliability Centered Maintenance model
Social • Equal distribution • Improved availability of social electricity services	There is a need to improve electricity service toward social benefit goals at household and community level like schools and health centers but also administrative and cultural centers. Knowledge communication is highly important to ensure users and services provider understand each other	This crosscutting area is involved in all the developed research projects

Alternatives for the Project Sustainability

Based on the results shown in the previous section, Centrosur has been looking to deepen the understanding of sustainability of the Yantsa Ii Etsari project, and strategic alliances were set up with the academia. For instance, the University of Cuenca and the Center for Development Research (ZEF), University of Bonn (Germany) have been collaborating through the development of undergraduate and graduate research projects in thematic areas linked with the above discussions on sustainability challenges. The results of these research experiences are presented in this section.

Local Utility Involvement and Stakeholders' Participation

Introduction

In 2010, when Centrosur began the Yantsa Ii Etsari project in the Amazonian province of Morona Santiago, it was necessary implementing internal changes in the structure of the company. For nearly 60 years of institutional life, the subject of the company was the distribution and commercialization of electricity through substations, distribution networks, transformers, meters, and so on, in short, through physical infrastructure that interconnects sources of electric generation "directly" with the end user.

In a joint effort between CONELEC and Centrosur, the project' scope was defined in isolated areas of Morona Santiago province within the concession area of the distribution company. It was initially estimated 2300 families to supply with electricity from SHS, which represented a new technology and way to provide electricity for both the company and the inhabitants of the area.

The first action consisted in creating a specific department within the company for carrying out the project implementation. The Renewable Energy Unit (UER) was conformed at the beginning by four electrical engineers who were trained to learn about experiences and status of various renewable energy projects developed in the country. During 2009 and 2010, the UER visited the communities in areas of difficult access where it was impossible to access to conventional electricity through networks. The main task of the UER was to get inputs for conducting a technical study about the best alternative of electrification. Also during this period, regulations for equipment, model contracts and community agreements, contract service provision, and the tariff were established.

In order to carry out the project, Centrosur defined the following stages of implementation:

- Preliminary survey to determine the current situation of the community
- Establishment of a community electrification committee
- Preparation of technical study
- Financing management

- Acquisition of equipment (technical specifications)
- Installation contract
- Transportation and equipment warehousing in the community
- Installation
- Training to the community on SHS management (by the contractor)
- Contract settlement
- Customer follow-up (by the electricity distributor)

The Yantsa Ii Etsari project had its first systems set up from January to June 2011. Two hundred ninety (290) SHS were installed in the parish Seville Don Bosco, Canton Morona. By July 2012, when the installation of the second phase finished, mostly in Canton Taisha, there was a significant block of new customers with SHS, totaling 2063. In subsequent years, the number of new facilities continued to increase (Table 7.2).

Centrosur's commitment is to ensure the service through the contract, which makes the user a regulated customer. Therefore, the follow-up performed by the company is essential for continuous operation and maintenance of SHS. During field visits, the work of the electrification committee, formed by community members, is verified, and the company supports the activities of this new actor. Knowledge and management of administrative and technical operators are strengthened as well as the concepts of minor maintenance and care for each SHS.

Relationship Between the Beneficiaries and the Electric Company

Centrosur looked for a model for sustainable rural electrification, where the community would be in close relationship to the company. The model should define the company management as a very strong influence to achieve institutional sustainability in coordination with community organizations. This institutional dimension is one of the five sustainability dimensions shown in Table 7.4.

Institutional sustainability refers to the organizational structures and processes that influence the success of the project within the local community. The stakeholders of this dimension include not only the distribution company and beneficiaries but also government officials such as mayors and parish presidents, opinion leaders as teachers, priests, and doctors, as well as traditional authorities and local associations represented by its president and trustee of each community. Even though this relationship worked before inclusion of the electrification project, the distribution company proposes a new scheme by introducing an additional actor: the electrification committee.

This scheme of strategic bridging (Garcia and Vredenburg 2003), between Centrosur and the user, proposes a new joint working relationship with the beneficiary community, behavior that in the past had only been present during the implementation stage of a project. A traditional paradigm is thus broken: the way in which the distributor used to be the electricity provider. These external factors (dispersed customers with no access to conventional network) are influencing changes that the distribution company must assume.

At this point, it is necessary to identify the most important changes that the electricity company had to make to face the Yantsa Ii Etsari project, in order to maintain its commitment to service and acceptance of the community' customers:

- Creating the Renewable Energy Unit (UER), a working group in charge of SHS projects
- Including in the training plan of the company topics such as renewable energy, community work, safety, and first aid in the Amazon
- Changes in the commercialization system of the company and creation of the residential photovoltaic rate (RF)
- Standardization of SHS equipment and their inclusion in the list of materials available at the company
- Creating a specific service contract for the service with SHS
- Creating regulation for the operation of the electricity committees and for administrative and technical staff

A Model for Sustainable Rural Electrification.

To promote sustainable rural electrification, first it is necessary to recognize the different dimensions of project sustainability, as shown in Fig. 7.2 and Table 7.4. These dimensions contribute to propose a model through three catalysts: a SHS design focused on the community, a sense of community ownership, and an active involvement of the distribution company.

SHS Design Focused on the Community

In order to design the most appropriate equipment for the community, one must know the kind of users that is intended to serve, their type of home, their habits (i.e., Shuar people are seminomadic), the economic income, service aspirations, etc.

Sense of Community Ownership

The understanding that the electrical service is possible through an SHS that uses a local resource gives to the community the feeling that it "owns" the project.

Distribution Company Involvement

The participation of the distribution company starts from identifying the community, supporting for the formation of the electrification committee and its operation, technical design, implementation of the project, and service customer management. The relationship between the sustainability dimensions and the catalysts is presented in Table 7.5 and Fig. 7.3.

Table 7.5 Dimensions and catalysts of a model for sustainable decentralized rural electrification

Catalyst / Dimension	SHS design focused on the community	Sense of community ownership	Involvement of the distribution company
Economic	The design guarantees adequate service with the necessary investment. The equipment meets standards to operate in places where they are installed, which ensures lower maintenance costs	When customers meet the payment of the prescribed fee and take care of equipment (less maintenance costs)	When company obtains resources for project implementation as well as for operation, maintenance, and replacement. The task of tariff collection is also important
Institutional	The design allows the beneficiaries involvement. For example, cleaning the panel and acknowledging of messages at the regulator display. SHS becomes a "new actor" in people's life	From electrification committees who "represent" the distribution company in their communities and have the acceptance of other local authorities	Operation and implementation of created structures such as the electrification committee and its representativeness in the community. This is reinforced by the application of regulations, contracts, meetings, etc., which are activities that show the operability of the committees
Technical (includes environmental)	Design is based on standards that provide equipment reliability. Besides, preventive maintenance depends on the training given to both each user and the technical operator in order to face minor maintenance problems	When customers use adequately the SHS and care equipment and perform preventive maintenance	From system design, standardization, and maintenance that can provide through the technical operator or its own staff. In addition, replacement and removal of equipment are in charge of the distribution company
Social	The design can promote both an equal distribution of electricity and opportunities for family/productive activities	When users care their installed systems as they recognize that through them it is possible to have electricity service	The company is able to educate people on the use and care of the system as well as on the rights and obligations assumed by the service contract

Conclusion

The local utility involvement seems crucial for sustainable decentralized rural electrification in the Ecuadorian Amazon region context. In the Yantsa Ii Etsari project, what the distribution company does is managing legitimacy (Schuman 1995) through formal institutional mechanisms (committees, regulations, contracts) and informal mechanisms ("culture of payment," "sense of community ownership," etc.).

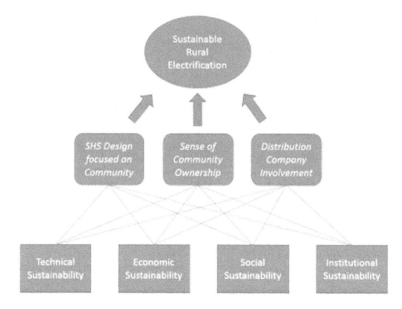

Fig. 7.3 Model of sustainable rural electrification. Source: Urdiales (2015)

Development of a Solar Boat Prototype

Introduction

The high level of isolation and lack of infrastructure makes accessibility to the Amazon region a very critical issue for development. It restricts people's mobility to long walks, riverboats, or small planes in order to reach markets and social services. Finding alternatives for transportation without endangering the Amazon ecosystem is an urgent challenge toward sustainability (Ordóñez and Guaman 2014). River transport in the Ecuadorian Amazon is one of the primary means of mobility in places where there are no roads (Jara-Alvear et al. 2013a, b). Rivers are used for navigation of people and goods in small boats with outboard motors. However, this mode of transportation causes significant environmental problems such as greenhouse gases – GHG emission – noise, fuel spills, and felling of large trees for manufacturing canoes (Ordóñez and Guaman 2014).

These problems can be mitigated by replacing gasoline outboard motors by electric propulsion systems. The application of these systems is not new. In 1839 Moritz von Jacobi Herman built one of the first electric boats (Morachevskii 2001), and today these systems have made significant progress worldwide as demonstrated by transatlantic crossings. Moreover, in 2013, the first Ecuadorian electric-solar boat was built and demonstrated its usability in Galapagos Islands, Ecuador, as a means for environmental education and sustainable tourism (Jara-Alvear et al. 2013a, b).

7 Sustainability of Renewable Energy Projects in the Amazonian Region 119

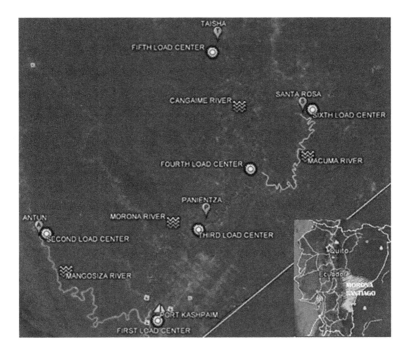

Fig. 7.4 Map of the study area. Souce: Guamán et al. (2015)

Ordóñez and Guaman (2014) developed a techno-economic study for replacing traditional outboard motors for electric outboard motors coupled with solar energy systems to recharge batteries within the electric propulsion systems. Their study aimed to assess the feasibility of a solar canoe adapted to the Amazon rivers.

Field Research and Data Collection

By the end of 2014, the Yantsa Ii Etsari project had 500 indigenous families settled on four riverbanks. A field trip along these four rivers was made for data collection on travel conditions in order to determine the design parameters of solar canoes (Fig. 7.4). The main origin and travel destinations were identified as well as the average speed and travel time for existing canoes (Table 7.6). It is important to note that routes 1–4, 2–5, and 3–6 represent a round trip, whereas route 7 represents the distance between two charging points, with an average travel time of 3 h.

Currently, 45 canoes provide transportation service along the above routes all year around. Based on 25 interviews, it was found that typical length of canoes is 12 m, they have an average load capacity of 1400 kg, and the most traditional equipment is a 13-horsepower (HP) outboard motor (Fig. 7.5).

Table 7.6 Main river routes, Morona Santiago, Ecuador

Route number	From	To	Name of river	Time (h)	Average speed (km/h)	Distance (km)
1	Kashpaim	Antun	Morona-Mangosiza	4	9	42
2		Taisha	Morona-Cangaime	9	9	92
3		Santa Rosa	Morona-Macuma	9	9	90
4	Antun	Kashpaim	Mangosiza-Morona	4	11	42
5	Taisha		Cangaime-Morona	9	11	92
6	Santa Rosa		Macuma-Morona	9	11	90
7 (charge center)	Kashpaim	Panientza	Morona	3	9	30

Source: Guamán et al. (2015)

Fig. 7.5 (Left) Traditional canoes, (right) traditional outboard motor "peque-peque". Source: Ordóñez and Guaman (2014)

Design of a Solar Canoe for the Amazon

The configuration for the electric-solar boat is based on the design of Jara-Alvear et al. (2013a), since it has demonstrated the technical capacity to displace up to 4000 kg in sea conditions. The configuration includes an electric outboard motor which is responsible to transform electricity into mechanical power to displace the boat, electrochemical batteries that are responsible to storage and provide the energy required by the electric outboard, and a photovoltaic system (onboard or onshore) which is responsible to recharge the batteries (Fig. 7.6).

The sizing of solar canoes was adapted from Jara-Alvear et al. (2013a) for displacement-type boats and includes seven steps:

Step 1: Define the design speed.
Step 2: Estimate the resistance for propulsion.
Step 3: Estimate propeller power.
Step 4: Estimate electric power needed.

Step 5: Estimate energy consumption.
Step 6: Size the battery bank.
Step 7: Size the photovoltaic generator.

Results: Configuration and Economic Analysis

Based on the sizing steps and using the data collected, Table 7.7 shows the resulting configuration of an electric 12-m canoe to cover the travel distance described in Table 7.6. The results show that the electric energy needs to cover the travel distances will require an area of solar panels bigger than the available on the canoes. Therefore, solar recharge stations should be placed on strategic locations along the travel route.

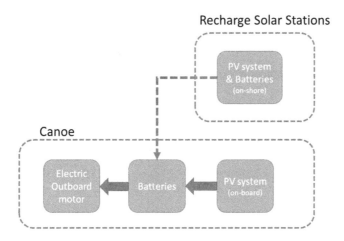

Fig. 7.6 Design concept solar canoes. Source: adapted from (Jara-Alvear et al. 2013a)

Table 7.7 Electric 12-m canoe configuration for each route

Route number	Power electric outboard (kW)	Capacity battery bank (kWh)		Solar generation (PVS)		Total weight of equipment (Ton) PVS and boat are not considered	
		La[a]	Li[b]	kWp (required)	kWp (onboard)	La	Li
1	4	18	13.42	4.3	1.61	0.39	0.14
3	4	42	26.85	9.8	1.61	0.87	0.28
7	4	15	10.74	3.3	1.61	0.33	0.12

[a]La: lead acid battery
[b]Li: lithium battery
Source: Ordóñez and Guaman (2014)

Table 7.8 Estimated cost for traditional vs. solar-electric 12-m canoe

	Route 1	Route 3	Route 7
Time of traveling (h)	4	9	3
Traditional canoe			
Investment outboard motor 13 hp. every 6 years ($)	$1030.00	$1030.00	$1030.00
Fuel ($/year)	$1929.78	$4342.00	$1447.33
Canoe investment every 4 years ($)	$600.00	$600.00	$600.00
Maintenance ($/year)	$200.00	$200.00	$200.00
Estimated annual reduction of CO_2 2.38 kg of CO_2 per liter of gasoline	8.29 ton	18.65 ton	6.21 ton
Solar-electric canoe			
Investment outboard motor 4 kW ($)	$3809.00	$3809.00	$3809.00
Investment 12 m fiberglass canoe ($)	$4000.00	$4000.00	$4000.00
Cost of Li-battery every 9 years ($)	$13,225.00	$26,450.00	$10,580.00
Cost of La-battery every 4 years ($)	$1623.60	$3788.40	$1353.00
Electricity cost—PV system ($)	$14,375.00	$31,050.00	$10,750.00
Maintenance ($/year)	$200.00	$200.00	$200.00

Source: Ordóñez and Guaman (2014)

Table 7.9 Economic feasibility of solar canoes

Results	Route 1	Route 3	Route 7
NPV (traditional canoe)	$31,228.00	$61,289.60	$25,215.61
NPV (solar canoe with La-Battery)	$28,277.89	$52,000.07	$23,772.00
NPV (solar canoe with Li-Battery)	$49,320.88	$92,324.24	$40,430.21
Payback period (traditional canoe)	11.6 years	>30 years	6 years
Payback period (solar canoe, Li-battery)	12 years	18.9 years	7.6 years
Payback period (solar canoe, La-battery)	6 years	8.5 years	4.1 years

Source: Guamán et al. (2015)

The weight and space onboard is a critical factor during the design phase. Considering that batteries are the heaviest and biggest component (see Table 7.7), the selection of lead-acid (La) or lithium (Li) batteries is a key aspect for the techno-economic study. La batteries are cheaper but heaver and less efficient, while Li batteries are lighter and more efficient but expensive.

The investment, operation, and maintenance cost for gasoline and electric outboard motors were estimated from interviews with canoe owners and literature review (Table 7.8). Using this information, the net present value (NPV) and payback period were estimated using an interest rate of 5% for a 20-year period (Table 7.9).

Conclusions

The selection of the best alternative depends mainly on the cost of the batteries, the type of boat, and the route to cover. The proposal would be economically viable for any battery technology in fiberglass boats with navigation time between 3 and 4 h.

From the economic point of view, the lead-acid battery is the best option but, from a technical viewpoint, lithium-battery is better since it has higher energy density (kWh/kg). This reduces weight and onboard space.

For the routes with travel times of 4 and 3 h, it is possible to navigate with the boat configuration presented in Table 7.7, and recharge centers should be located at the beginning and end of the routes. For longer travel times (i.e., 9 h) it is recommended to have recharge centers in strategic locations along the routes (Fig. 7.4). This will reduce the number of batteries and solar photovoltaic panels required for autonomy, reduce weight on board increasing space for more cargo and passenger loads, and reduce propeller power consumption (lower system cost).

The limitations for electric boats are the high up-front cost of technology, reduced travel autonomy, and low speed, all limited mainly by existing storage technology (Del Pizzo et al. 2010). Nevertheless, coupling electric boats with onboard renewable energy generation and charging stations along the travel route could potentially help to surpass these barriers and facilitate their adoption in isolated areas of the Amazon region.

The electric-solar boats might also serve as a means of transport to Centrosur technicians so they can follow up the installed SHS. More important, along the analyzed routes there are communities' beneficiaries of the Yantsa Ii Etsari Project; thus, the electric-solar boats will also provide a safe and clean way of transport for people in the region. This research project shows that electric-solar canoes provide not only techno-economic advantages but also social and environmental benefits, becoming a sustainable alternative for the Yantsa Ii Etsari project in particular and for river transport in the Ecuadorian Amazon region in general.

Design of an Automatic Reliability-Centered Maintenance Model

Introduction

From maintenance reports of the project Yantsa Ii Etsari, it was found that 32.7% of the SHS have failed in the period 2012–2014 for some technical reason (Fig. 7.7). Compact fluorescent lamps (CFLs), regulators, and inverters are the components with the highest failure rate (Table 7.10). Centrosur is enforced to provide a good quality electricity service with a highly subsidized electric tariff (user's fee payment is $17.56/year). This amount is not enough to cover operation and maintenance of SHS, and it makes the project dependent on external funding to be sustainable in the long term. In addition, a high failure rate of equipment combined with the absence of a waste management system in Amazon communities threatens the ecosystem, since toxic substances from electronic waste could be released to the environment, for instance, mercury, which is contained in CFLs.

A reliability-centered maintenance (RCM), which started in the airline industry in the 1960s, is a structured framework that helps to improve maintenance decision through the analysis of functions and potential failures of physical assets (e.g., SHS) and schedule maintenance task in order to enhance reliability at the lowest cost

Fig. 7.7 Failures as compared to total installed SHS. Source: Urdiales (2015)

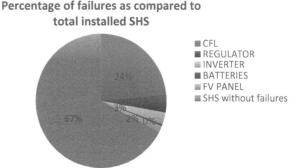

Table 7.10 Failure rate of SHS, period 2012–2014

Equipment	Frequency of failure	% Share
Compact fluorescent lamps (CFLs)	683	72.05
Regulator	233	12.87
Inverter	84	8.86
Solar panel	47	4.96
Battery	12	1.26
Total	948	100

Source: Urdiales (2015)

(Orellana and Porras 2014). The standard SAE JA101 establishes the minimum criteria to implement a RCM, which has to systematically answer the questions presented in Table 7.8. In order to assist decision-makers on implementing a RCM strategy, Orellana and Porras (2014) developed an automated system that could help to answer such questions in the context of the project Yantsa Ii Etsari (Table 7.11).

Research Approach and Results

The research approach included three phases. First, data collection of written maintenance reports was reviewed, and a household survey in 25 communities was conducted for field investigation of SHS status and users' experience. Second, FF, FM, FE, and FC were identified using criticality analysis (Moss and Woodhouse 1999), root cause analysis (RCA), and failure mode and effects analysis (FMEA). Third, based on this analysis, an entity-relationship (ER) model was implemented to facilitate the execution of RCM in Centrosur.

Figure 7.8 shows the resulted ER that has the following purposes in order to assist for the implementation of a RCM strategy:

1. Functional diagram of the SHS and its components, including principal and secondary functions.
2. Register FMEA results, which include system and component's function, operational context, FM, FE, and FC.

3. Calculation of the reliability index called Risk Priority Index (IPR) in order to prioritize the most critical components in terms of safety, environmental, and operational security.
4. Define and register corrective, preventive, and predictive maintenance tasks.
5. Elaborate maintenance reports and work orders for maintenance staff.

Entities of the model (i.e. users, SHS components, FM, FE, and FC) have associated attributes that provide further relevant information for RCM and maintenance decisions. For instance, user's name, code, and geographical location are key to connect with other database of Centrosur. SHS were disaggregated in their different components in order to define FF, FM, FE, and FC, which are used to perform a FMEA. The results are the IPR for each SHS component (Orellana and Porras 2014). This provides the basis to plan and prioritize maintenance tasks (preventive, corrective, predictive).

The implementation of the ER model was done in Microsoft Access, where entities tables were connected (Fig. 7.9) and an interface was designed for each table in order to facilitate data input and also retrieve information by Centrosur staff at administrative and technical level (Fig. 7.10). In addition, this tool facilitates the elaboration of reports about the status of the installations and work orders for maintenance staff (Fig. 7.11).

Centrosur staff checked and confirmed the usability of this tool. For instance, Table 7.12 shows the resulted IPR for all SHS components of Yantsa Ii Etsari. It provided valuable information to focus maintenance efforts. For example, it was found that solar regulator is the most critical component though it has an acceptable frequency of failure (Table 7.10). A regulator controls the state of charge of the battery, which has a limited lifetime and requires periodic replacement (3–4 years). If batteries have frequent and deep discharges due to regulator failures, its lifetime is highly affected, and the time for replacement could be reduced considerably (1 year or less). Therefore, it will have negative effects on planned maintenance cost

Table 7.11 RCM framework and questions

Criteria	Question in the context of SHS
Functions	1. What are the functions and desired performance of SHS in the Amazon?
Functional failures (FF)	2. What are the functional failures that might occur that prevent SHS to perform its expected function?
Failures modes (FM)	3. What are the events that likely cause a functional failure in SHS?
Failure effects (FE)	4. What happens when a functional failure occur in SHS?
Failure consequences (FC)	5. In what way does each SHS's functional failure matter in terms of safety, environment, operational, and nonoperational consequences?
Planning proactive tasks	6. What can and/or should be done to predict or prevent SHS failure?
Default actions	7. What should be done if a suitable proactive task cannot be determined?

Source: Orellana and Porras (2014)

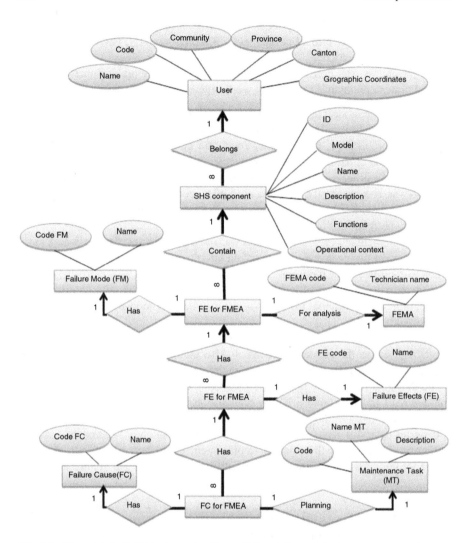

Fig. 7.8 ER model for RCM in the project Yantsa Ii Etsari. Source: Orellana and Porras (2014)

and the production of an unexpected waste. Consequently, regulators are key elements for maintenance and SHS reliability.

Conclusion

This research work proposes an automatic model to assist in the implementation of RCM strategies for rural electrification projects. The purpose of this tool is enhancing system reliability at the lowest cost, through the analysis of likely failures, their causes, effects, and consequences on safety, environment, and budget, to improve plan maintenance tasks and their frequency.

7 Sustainability of Renewable Energy Projects in the Amazonian Region

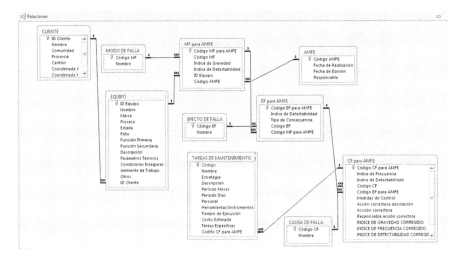

Fig. 7.9 Database construction of the RCM model. Source: Orellana and Porras (2014)

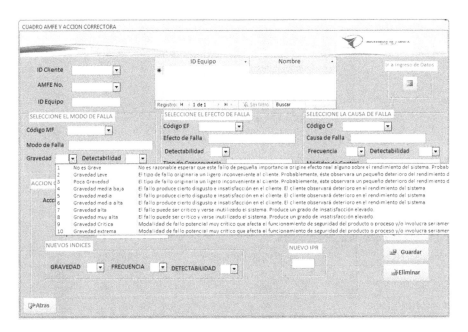

Fig. 7.10 Interface example of RCM model, FMEA windows

High reliability power systems usually mean more investment on high quality and robust equipment. This could affect the already high cost of SHS. However, in the long term, it could have an effect on reducing operation and maintenance costs, especially in the Amazon region context where accessibility is restricted and expensive. Moreover, SHS reliability could enhance components lifetime and therefore

Fig. 7.11 Automatic report, work order. Source: Orellana and Porras (2014)

Table 7.12 Critical elements of a SHS

Components	Criticality index	Ranking
Regulator	164	1
Inverter	120	2
Lamps (CFL)	96	3
Battery	45	4
Solar panel	45	5
Battery fuse	58	6
Loads fuse	29	7
Electric kit installations (cables, interrupters)	29	8

Source: Orellana and Porras (2014)

reducing the production of electronic waste and the risk to release toxic substance in the ecosystem. However, a further research on RCM is needed in order to understand both the optimal levels of reliability and its cost-benefit in the Amazon region.

Integrated Waste Management System

Introduction

This section summarizes the proposal for an integrated waste management generated in the stages of pre-installation, installation, operation, and abandonment of the Yantsa Ii Etsari project developed by Urdiales (2014). The main objectives of this research were:

- Define the baseline of the study area, determining the environmental factors that may be affected by project activities.
- Identify, evaluate, and categorize potential environmental impacts.
- Formulate an environmental management plan (EMP) for each stage of the project.

Baseline Definition

After obtaining the main environmental factors, the project activities that could affect these factors were identified through field visits, photographs, interviews, and conversations with Centrosur staff and SHS users. This allowed developing flowcharts for each project process: pre-installation, installation, operation, and abandonment. Each process considers inputs, activities, outputs, and waste/emissions. Being a project with less than 5 years of operation, abandonment activities were not considered. The flowcharts are presented next.

Pre-installation Flowchart

In terms of environmental impact of this process, the only activity considered is "working test" whose input, output, and waste are presented in Table 7.13. The other activities of the process, purchasing, receiving, offloading, and warehousing, do not generate significant environmental impacts (Fig. 7.12).

Installation Flowchart

Most activities of this process generate environmental impacts, mainly waste and/or emissions, as shown in Table 7.14 that represents the flowchart of the process. Figure 7.13 shows the main activities of the installation process.

Table 7.13 Flowchart of pre-installation

Input	Activity	Output	Waste/emissions	Note
Components of the off-loaded SHS	Working test	Components of the tested SHS	Defective components or in poor conditions (hazardous solid waste)	

Source: Urdiales (2014)

Fig. 7.12 Receiving equipment (left) and working tests (right). Source: Urdiales 2014

Operation Flowchart

The most significant environmental impact of this process has to do with the corrective maintenance of the systems, particularly with the activities of "replacement of defective parts" and "transport" (Table 7.15). The other activities related to preventive maintenance (cleaning panels, installation checking, etc.) do not generate impacts (see Fig. 7.14).

Identification and Significance of Environmental Impacts

In year 2014, a survey to 65 households out of 2060 was conducted for obtaining necessary information from the environmental aspects, strengths, and weaknesses regarding the solid waste management of SHS in the communities. The main results of the survey were the following (Urdiales 2014).

All respondents knew the utility of SHS, and 72% said they have basic knowledge on how to operate and do basic maintenance; 45% of respondents did not want to connect any additional devices to their SHS, and the remaining want to buy and connect new equipment that could surpass SHS capacity.

More than half of respondents (56%) said that they did not face problems with SHS; the remaining 44% had maintenance/operation problems. When the problems are not minor, Centrosur solves them in a maximum of 4 months. The components with the higher failure rate are CFLs (Table 7.10). From survey, 77% said they know

Table 7.14 Flowchart of installation

Input	Activity	Output	Waste/emissions	Note
Components of packed SHS	Transportation from warehouse	Components of transported SHS	Air emissions by type of transport. Domestic solid waste	
Components of transported SHS	Offloading in destination	Off-loaded SHS in final destination	Plastics, cardboard, etc. (industrial nonhazardous solid waste)	
Off-loaded SHS in final destination	Excavation of poles	Erect poles	Snatches of poles, debris (industrial nonhazardous solid waste)	Installation outside housing
Erect poles, panels	Location of panels	Panels located and fastened	Packaging (plastic, cardboard), trees, debris	Installation outside housing
Panels located and fastened, battery	Battery connection	Connected battery		Installation inside housing
Connected battery	Equipment connection	Connected equipment	Scraps of metal, wires, and tapes (industrial nonhazardous solid waste)	Installation inside housing
Connected equipment	Cabling	Installed equipment	Snatches of cables, tapes, etc. (industrial nonhazardous solid waste)	Installation inside housing

Source: Urdiales (2014)

how to dispose the lamps according to instructions until the new equipment arrive for replacement. Regarding batteries, which will require periodic replacements, 64% of respondents did not know how to dispose them.

About environmental impacts of SHS, 100% of respondents asserted that there is no noticeable pollution affecting water, air, and soil factors in their communities; 76% believe that the installation and use of SHS generate a positive impact on their daily activities. In addition, 59% said that the installation and use of SHS generate a positive visual impact to the environment, while 30% did not reply, and 11% said that this impact is not positive.

In order to contrast survey results, environmental impact identification and assessment were performed using a double-entry matrix (Leopold matrix), which in one axis includes the main phases of project activity, while the other axis includes the environmental factors.

The main negative environmental impacts are (Urdiales 2014):

- Air emissions from internal combustion engines of aircrafts, vehicles, and boats, in which the components of SHS are transported to the communities
- Potential degradation of soil quality during both offloading of equipment and digging of poles

Fig. 7.13 Transportation and offloading of equipment. Source: Urdiales (2014)

Table 7.15 Flowchart of operation

Input	Activity	Output	Waste/emissions	Note
Spare parts	Replacement of defective parts	Components with parts in good condition	Defective components or in poor conditions	Corrective maintenance
Defective parts, packing	Transport	Transported parts or components	Plastic, cardboard, etc. (nonhazardous solid waste)	

Source: Urdiales (2014)

- Alteration of vegetation and natural habitats during transportation by river or land and offloading, as some of the communities are within fragile ecosystems
- Proliferation of insects and weed invasion in warehousing activities as well as in external and internal connections of equipment and their final disposal
- Change on cultural models (customs) in activities of transport, installation, cleaning, and washing of panels when preventive maintenance, replacement of damaged parts, and/or final waste disposal are performed
- Possible impacts on the population health by faulty or no final waste disposal when the components are changed (batteries, CFLs, etc.)

After building the matrix of (negative) impact significance, the conclusions were:

- Irrelevant impacts (54%): minimal deterioration of the landscape when SHS are transported and installed; affectation to bugs species and/or archeology when

Fig. 7.14 Electrified houses of the Yantsa Ii Etsari project. Source: Urdiales (2014)

excavating for poles; invasion of weeds, insects, and vectors in the system components when there is no proper cleaning
- Moderate impacts (34%): disturbance to natural vegetation, wild species, and protected areas during the processes of transport of materials by the increased use of people of trails to access to communities; air emissions depending on whether the transport is by river or air; potential impact on the health of workers that handle components in both transport and installation
- Severe/critical impacts (9%): solid waste generation (hazardous and nonhazardous) during the different stages of the project

Finally, an important positive impact is the source of employment for local people as canoeists, stevedores, and installers. Both electricity itself and employment contribute to local sustainable development.

Environmental Management Plan (EMP)

Once you have identified and assessed the environmental impacts of the project in its different stages, it is necessary to undertake an "action plan" or EMP. A project's environmental management plan (EMP) consists of the set of mitigation, monitoring, and institutional measures to be taken during implementation and operation to eliminate adverse environmental and social impacts, offset them, or reduce them to acceptable levels. The plan also includes the actions needed to implement these measures" (WB 1999).

The EMP will allow an optimal integration between different processes of the photovoltaic project and environmental factors identified in the area of influence. This Plan should be understood as a dynamic tool and therefore variable over time. This means that it should act as a continuous improvement mechanism on the project's environmental aspects and their impacts.

The proposed EMP for the project Yantsa Ii Etsari contains the following programs, each with its specific activities or subprogram, schedule, budget, and responsible people (Urdiales 2014):

- Prevention and mitigation
- Occupational safety and health
- Contingencies and risks
- Waste management
- Community relationships
- Abandonment and restitution of the area
- Monitoring, control, and environmental follow-up
- Training, education, and diffusion

Since the production of solid waste (hazardous and nonhazardous) at different stages of the project has been categorized as severe/critical impact, it is important to tackle this problem holistically. Improper handling of waste causes contamination of water, air, and soil, landscape deterioration of the area of influence, as well as possible effects on people health.

Toward an Integrated Solid Waste Management

Even though Centrosur has both an instructive for handling materials and wastes (Code I – DIGARS-349) and specific forms for handling materials and waste during the construction, operation, and maintenance of SHS, a deeper analysis that allows integrating the impacts identified and the measures presented in the EMP is necessary (Urdiales 2014).

Integrated solid waste management is ruled by the current legislation regarding the control and waste management, summarized in laws and regulations issued mainly by the Ministry of Environment. Besides, the Ecuadorian Standard Construction NEC-10 Part 14.02 (Renewable Energy) Generation Systems with Photovoltaic Solar Energy for Isolated Systems and Network Connection (up to 100 kW) in Ecuador is taken into consideration (INEN 2010).

Six stages in the integrated waste management proposal were established, from the waste identification at the source to final disposal. They are (Urdiales 2014):

- Primary disposal
- Transport
- Secondary disposal
- Classification
- Treatment
- Final disposal

Each stage has clearly defined those responsible for its implementation, depending on the different players who develop the project (contract managers, contractors, technicians, staff, etc.)

The main wastes produced in the Yantsa Ii Etsari project are (Urdiales 2014):

- Domestic solid waste: due to food consumption, remains of packaging paper, plastic, cardboard, and other inert inputs.
- Industrial solid waste: remains of construction materials from assembly and disassembly activities. These residues are classified as:

- Nonhazardous industrial solid waste: products from disassembly of equipment such as uncontaminated scrap
- Hazardous industrial solid waste: batteries, panels, CFLs, and other components that contain heavy metals that can pollute or create pollution risks

Considering both the EMP (waste management program) and Centrosur's instructive for handling materials and wastes, two forms that allow a project integrated waste management were developed:

1. Management of domestic solid waste and nonhazardous industrial solid waste
2. Management of hazardous industrial solid waste

Conclusion

The integrated waste management of the Yantsa Ii Etsari project is an important step toward environmental sustainability. The effective implementation of the EMP will help to prevent and control all negative impacts as well as socializing the project and encouraging environmental responsibility in beneficiary communities.

Sustainability Assessment of Rural Electrification: Further Steps

From previous results and as part of a research in progress carried out by ZEF-University of Bonn in cooperation with Centrosur, sustainability assessment protocols of decentralized rural electrification in the Amazon region should include the following aspects.

Environmental Dimension

Weigh the environmental impacts of the electricity distribution with renewable energy on the Amazon ecosystem from planning, installation, maintenance, and decommissioning phases. A key aspect to take into account is waste production and management. In addition, the side effects of electricity uses in fragile ecosystems are critical as they could foment environmental degradation and deforestation (e.g., sawmills). Finally, the contribution of decentralized rural electrification on mitigation and adaptation to climate change is also a global concern to take into account.

Economic Dimension

Appraise the financial and economic viability of deploying renewable energy technology in rural and low-income areas. It is perhaps the most studied issue on rural electrification research. However, deepening the understanding on cost-benefit analysis of electricity access on social improvements and environment protections is an area that could shed light on the externalities to take into account in financial analysis of projects.

Institutional Dimension

Institutions are the key drivers for development and stakeholder engagement; therefore, sustainability models should appraise stakeholder capacity and engagement but also participation on projects development.

Technical Dimension

Appraise availability, reliability, and affordability of technology for the Amazon region context. Perhaps renewable energy is not affordable for low-income people, but it is the only solution available today to provide electricity to scattered households with low energy demand. Focusing the analysis on reliability and meeting users demand for development are key areas for decision-making from the technical viewpoint.

Social Dimension

It is perhaps the most complicated dimension to assess, since it is dependent and influenced by development as a whole, and not only on the provision of electricity service (Ilskog 2008). However, assessing the social benefits and equity issues are highly relevant for project evaluation and monitoring. This crosscutting area is involved in all the previous dimensions. Some of the social key issues are:

- The impact of electricity on income generation
- Support for communal services (health centers, educational centers, etc.)
- Preservation of traditional knowledge and practices of the cultural diverse indigenous population of the Amazon
- Participation and integration of all relevant stakeholders in the process of rural electrification

Spatial Dimension of Sustainability

Isolation has been the main physical barrier to boost socioeconomic development in the Amazon region and other parts of the world. Perhaps this isolation has protected the Amazon from urbanization and agriculture expansion. Moreover, the spatial distribution of different types of ecosystems and indigenous people living in different levels of socioeconomic development create a diverse range of human-environment interactions. In this regard, making explicit this spatial variation could expand the level of analysis but overall the communication and participation of stakeholders in order to progress toward sustainable energy development in fragile ecosystems. For instance, maps that shows how electricity access has improved accessibility through the reduction of travel time and cost to reach health and education services will have an impact on people development but also on reducing greenhouse gas emissions from transportation (air, car, boats).

Stakeholders' Participation

The multidimensionality and fuzziness of sustainability concepts demand the understanding and integration of the different viewpoints. Table 7.4 was an initial effort in that direction. However, any sustainability assessment model should consider the active participation of different stakeholders from government, indigenous communities, and civil society that have an interest or influence on decentralized rural electrification. Their knowledge and needs will make any protocol or model for sustainability assessment relevant and salient for policymaking.

In order to monitor and guide the transition toward sustainability in rural electrification projects, clear objectives have to be defined and assessed. However, measuring sustainability is a challenge that has mobilized the scientific community to propose a diversity of tools to support decision-makers. Ness et al. (2007) provide a categorization of existing tools classified in three big groups: indicators, product-related assessment, and integrated assessment methods.

- Indicators are extensively used for sustainability assessment. In this chapter, by using an existing sustainability framework of indicators (Ilskog 2008), a research plan was proposed in order to shed light on how to promote sustainability of rural electrification projects.
- Product-related assessment focuses on assessing flows of resources and services. For instance, a "Life Cycle Assessment" of SHS could help to understand the contribution of rural electrification in greenhouse emissions scenarios. In addition, a "product material flow analysis" could help to assess the material input per unit of energy delivered, which is a key aspect on waste management.
- Integrated assessment looks at system analysis approaches. It means that rural electrification should not be treated as a technical or economic problem, but as an interconnected entity. This type of analysis is more interested on relationship rather than elements of the system. For instance, "system dynamics" provide a potential platform for research on modeling rural electrification as a system, where social, environmental, and technical variables are integrated and simulated.

Conclusions and Discussion

The ecological importance of the Amazon region for climate regulation but also as a source of high biological and cultural diversity is generally accepted. Human development in this region is very low, and improving their living conditions could help to counteract the environmental degradation of this sensible ecosystem. Providing electricity service sets the background for the provision of other services like health, education, and telecommunication. However, deploying technology for electricity supply in isolated and low-income areas represents a challenge still unsolved worldwide.

This chapter introduces the project Yantsa Ii Etsari, which has almost reach universal access to electricity in the southern Ecuadorian Amazon. In the sake to improve the sustainability of electricity provision, a cooperation with the academia of Ecuador (University of Cuenca) and Germany (ZEF-University of Bonn) was set in order to research on key areas that could provide better understanding on how to improve the sustainability of the project.

Four research projects were presented "local utility involvement and stakeholder participation," "development of a solar boat prototype," "design of an automatic reliability-centered maintenance model," and "integrated waste management system." Each focused on different sustainability dimensions and provided the basis for discussion and policy.

This chapter demonstrates that cooperation between practitioners (Electricity Company) and researchers (University) provides a unique opportunity to discuss and understand how to promote sustainability in isolated and fragile areas. However, how to translate research outcomes into energy policies is a challenge.

References

ARCONEL (2013) Plan Maestro de Electrificacion Rural 2013–2022. Available at: http://www.conelec.gob.ec/contenido.php?cd=10329&l=1

Brent A, Rogers D (2010) Renewable rural electrification: sustainability assessment of mini-hybrid off-grid technological systems in the African context. Renew Energy 35(1):257–265. Available at: http://linkinghub.elsevier.com/retrieve/pii/S0960148109001499 Accessed: 2 Apr 2012

CODENPE (2012) Map of Indigenous Natinalities of Ecuador. Available at: http://www.codenpe.gob.ec/index.php?option=com_content&view=article&id=125&catid=96

Del Pizzo A, et al (2010) Design criteria of on-board propulsion for hybrid electric boats. In: 19th international conference on electrical machines, ICEM

Elkington J (1999) Cannibals with forks: triple bottom line of 21st century business. Capstone Publishing Ltd, Oxford

Garcia P, Vredenburg H (2003) Building corporate citizenship through strategic bridging in the oil and gas industry in latin America. J Corp Citizensh (10):37–49

Guaman F, Ordoñez J, Espinoza JL, Jara-Alvear J (2015) Electric-solar boats: an option for sustainable river transportation in the Ecuadorian Amazon. In: 6th International conference on energy and sustainability, WIT Transactions on Ecology and The Environment, Wessex, UK, Vol 195, pp 439–448

Ilskog E (2008) Indicators for assessment of rural electrification—an approach for the comparison of apples and pears. Energy Policy 36(7):2665–2673. Available at: http://linkinghub.elsevier.com/retrieve/pii/S0301421508001407. Accessed 25 July 2011

INEC (2010) Population and housing census of Ecuador. Available at http://www.ecuadorencifras.gob.ec/base-de-datos-censo-de-poblacion-y-vivienda-2010/

INEN (2010) Norma Técnica Ecuatoriana: Sistemas de generación con energía solar fotovoltaica para sistemas aislados y conexión a red de hasta 100kW en el Ecuador

Jara-alvear J, Urdiales L (2014) "Yatsa Ii Etsari" (Light from our Sun): Lessons for sustainable photovoltaic rural electrification on remote Amazon indigenous communities in Ecuador. In: Conference MES-BREG 2014. pp 3–6

Jara-Alvear J, Pastor H, et al (2013) Embarcaciones solares, una evolución al transporte marino en las islas Galápagos, Ecuador. In: 1st International congress and scientific expo ISEREE 2013. Available at http://www.iner.ec/congreso/images/documentos/Articulos/Posters/ISEREE2013_Jara_Jose.pdf

Jara-Alvear J, Callo-Concha D, Denich M (2013b) Decentralized rural electrification projects with solar home systems in the Ecuadorian Amazon: a holistic analysis. In: 1st International Congress and Scientific Expo ISEREE 2013. pp 1–8. Available at: http://www.iner.ec/congreso/images/documentos/Articulos/ISEREE2013_Jara-Alvear_Jose.pdf

Johansson T, Goldemberg J (2002). Energy for sustainable development (UNDP). Available at: http://www.undp.org/content/dam/aplaws/publication/en/publications/environment-energy/www-ee-library/sustainable-energy/energy-for-sustainable-development-a-policy-agenda/Energy%20for%20Sustainable%20Development-PolicyAgenda_2002.pdf

Mainali B, Silveira S (2015) Using a sustainability index to assess energy technologies for rural electrification. Renew Sust Energ Rev 41:1351–1365. https://doi.org/10.1016/j.rser.2014.09.018

Mebratu D (1998) Sustainability and sustainable development: historical and conceptual review. Environ Impact Assess Rev 18(6):493–520

Mendes WDS et al (2009) An outbreak of bat-transmitted human rabies in a village in the Brazilian Amazon. Rev Saude Publica 43(6):1075–1077

Morachevskii A (2001) Works of academician B.S. Jacobi in the field of applied chemistry. Russ J Appl Chem 74(8):1422–1425. Available at: http://download.springer.com/static/pdf/390/art%253A10.1023%252FA%253A1017426608956.pdf?auth66=1424703513_b1715639b1bf5cbfa40f2eee6673c289&ext=.pdf. Accessed 23 Feb 2015

Moss TR, Woodhouse J (1999) Criticality analysis revisited. Qual Reliab Eng Int 15(2):117–121

Ordóñez J, Guaman F (2014) Estudio de viabilidad tecnica, economica y ambiental de lanchas solares para transporte en los rios de la Amazonia sur del Ecuador. University of Cuenca

Ness B, Urbelpiirsalu E, Anderberg S, Olsson L (2007) Categorising tools for sustainability assessment. Ecol Econ 60(3):498–508. https://doi.org/10.1016/j.ecolecon.2006.07.023

Orellana E, Porras J (2014) Sistema automatizado de mantenimiento centrado en confiabilidad (MCC) para proyectos de electrificacion rural con sistemas individuales fotovoltaicos en la Amazonia Ecuatoriana. University of Cuenca. Available at: http://dspace.ucuenca.edu.ec/handle/123456789/20922

Schuman M (1995) Managing legitimacy: strategic and institutional approaches. Acad Manag 20(3):571–610

SENPLADES (2013) Plan nacional del Buen Vivir 2013–2017. Available at: http://www.senplades.gob.ec/c/document_library/get_file?uuid=5a31e2ff-5645-4027-acb8-6100b17bf049&groupId=18607

Urdiales D (2014) Gestion integral de residuos en etapas de preinstalacion, instalacion, funcionamiento y retiro de sistemas fotovoltaicos de la segunda etapa del proyecto Yatsa Ii Etsari en la Empresa Regional Centrosur C.A. University of Cuenca

Urdiales L (2015) Procedimiento para la electrificacion en zonas aisladas: caso canton Taisha, Morona Santiago. Available at: https://www.google.de/url?sa=t&rct=j&q=&esrc=s&source=web&cd=2&cad=rja&uact=8&ved=0ahUKEwjKtc_h2I3LAhVBJw8KHZwXApwQFggkMAE&url=http://dspace.ucuenca.edu.ec/bitstream/123456789/21428/1/Tesis.pdf&usg=AFQjCNFFq8PtCKr22XhHvfsXV6zak-4ITA&bvm=bv.114733917,d.ZWU

Vasconez J (2010) Informe general del inventario de Instalaciones Fotovoltaicas y otros sistemas de energias renovables de hasta 500 kW. Ministry of Electricity and Renewable Energy, Quito, Ecuador

World Bank (1999) Environmental management plan, OP 4.01 - Annex C, january. At https://policies.worldbank.org/sites/ppf3/PPFDocuments/3903Operational%20Manual%20-%20OP%204.pdf

WCED (1997) Report of the World Commission on Environment and development: our common future. Available at: http://www.un-documents.net/ocf-02.htm#I

Chapter 8
Estimation of Landfill Methane Generation from Solid Waste Management Options in the Galápagos Islands

Rodny Peñafiel, Lucila Pesántez, and Valeria Ochoa-Herrera

Introduction

Solid waste management in fragile environments is a serious problem especially on small isolated islands that are tourist attractions (Chen et al. 2005). Waste generation in such places has rapidly increased and waste management options are limited (Mohee et al. 2015). Furthermore, increasing tourism, lack of treatment or disposal infrastructure, and difficulty of exporting waste to the continent worsen the situation (Santamarta et al. 2014). Galapagos Islands have a unique and fragile environment threatened by inefficient solid waste management (Baine et al. 2007). Moreover, sanitation in these islands requires significant improvement in order to properly protect marine resources (Dirección del Parque Nacional Galápagos 2014). Solid waste final disposal in landfills has been primarily used in isolated environments. Nevertheless, recyclable waste separation and biomass composting are considered to have important environmental advantages over landfilling (Mohee et al. 2015). Methane emissions in landfills are of global concern due to their global warming potential (GWP). Landfills are import methane generators, representing approximately 19% of all anthropogenic emissions (Czepiel et al. 2003). Landfill gas may be collected in pipes, and its energy may be recovered for local use (Thompson et al. 2009); nevertheless in conventional small-scale landfills (less than two million tons) methane recovery rate may be fairly small, around 5% (Bo-Feng et al. 2014). A significant increase in population and solid waste generation is expected in the Islands for the next 15 years. The research reported in this chapter estimates landfill methane production from waste management disposal options for Santa Cruz, San Cristobal, and Isabela Islands. 90% of the Galapagos population lives on these three islands (INEC 2010). An evaluation of two waste management disposal scenarios in terms

R. Peñafiel • L. Pesántez • V. Ochoa-Herrera (✉)
Colegio de Ciencias e Ingenierías, Universidad San Francisco de Quito, Quito, Ecuador
e-mail: vochoa@usfq.edu.ec

© Springer International Publishing AG 2018
M.-E. Tyler (ed.), *Sustainable Energy Mix in Fragile Environments*,
Social and Ecological Interactions in the Galapagos Islands,
https://doi.org/10.1007/978-3-319-69399-6_8

of their landfill methane generation was conducted. This evaluation considered the following: (1) all waste generated in the islands will be landfilled and biogas will be vented; (2) 60% of recyclable waste will be segregated and exported to the mainland; (3) 40% of biomass will be composted, and the rest will be landfilled.

Estimation of Solid Waste Generation

Population growth in the islands was based on INEC[1] 2010 projections. Solid waste production was determined in accordance with MAE (2016)[2] solid waste estimations and records from the Environmental Department of San Cristobal municipality. Table 8.1 shows information on the population and solid waste generation from 2010 to 2014 in San Cristobal. A per capita solid waste generation rate between 0.41 and 0.61 kg capita^{-1} d^{-1} was determined. These results are in the expected range of solid waste generation rates for isolated islands. According to Mohee et al. (2015) in Caribbean and Pacific islands with a Human Development Index (HDI) similar to Ecuador, a solid waste generation rate between 0.4 and 2.5 kg capita^{-1} d^{-1} is expected. The solid waste generation rates in San Cristobal for the period 2016 to 2031 were estimated from records reported in that island. An annual increment of the solid waste generation rate of 3.5×10^{-2} kg capita^{-1} d^{-1} was determined. Due to similarities in socioeconomic and environmental conditions of the islands (INEC 2010), the same solid waste production rate and annual increment were chosen for Santa Cruz and Isabela.

Figure 8.1 shows the forecasted population growth for the period 2016–2031 according to INEC (2010) predictions.

Figure 8.1 shows a 1.4-fold population increase in Santa Cruz for the period 2016–2031. Figure 8.2 shows a 2.7-fold increase in solid waste generation for that island. On the other hand, in San Cristobal and Isabela islands, the corresponding increases in population were 1.3 and 1.4, and the increases in solid waste generation were 2.6 and 2.8, respectively. The accumulated solid waste generation quantities at

Table 8.1 Solid waste generation in San Cristobal

Year	San Cristobal population	Solid waste generation (ton year^{-1})	Solid waste generation rate (kg capita^{-1} d^{-1})
2010	7707	1157	0.41
2011	7899	1210	0.42
2012	8095	1361	0.46
2013	8293	1859	0.61
2014	8493	1763	0.57

---

[1] INEC (2010): *National Institute for Statistics and Census of Ecuador.*
[2] MAE: *Ministry of Environment of Ecuador.*

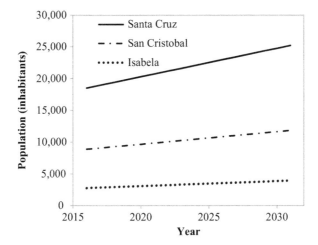

Fig. 8.1 Forecasted population 2016–2031 (based on INEC[1] projections)

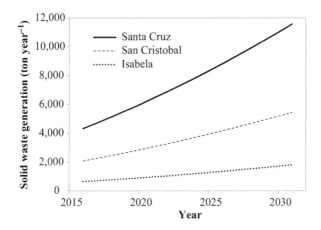

Fig. 8.2 Forecasted solid waste production 2016–2031

the end of the 15 year-period are 123,495 tons in Santa Cruz, 58,436 tons in San Cristobal, and 18,951 tons in Isabela.

Table 8.2 shows the solid waste composition of the three islands. In Santa Cruz and San Cristobal, the largest waste component was residual wastes (43% and 50%) and organic wastes (food residues) were the second largest contributor (38% to 39%). These results are similar to the solid waste characteristics from various Caribbean and Pacific islands, which show organic content ranging between 41% and 46% (Mohee et al. 2015). By comparison, the largest component in Isabela was organics (86%). This island has a much smaller population than Santa Cruz and San Cristobal and less tourism influence (INEC 2010), which may explain the composition differences.

Table 8.2 Solid waste composition in the Galapagos Islands

	Santa Cruz	San Cristobal	Isabela
Component	% (w/w)	% (w/w)	% (w/w)
Organics	39	38	86
Plastics, glass, and metals	11	9	6
Paper and cardboard	7	3	1
Rest	43	50	7

Estimation of Landfill Methane Generation

Two scenarios were considered in estimating landfill methane generation: (1) all waste will be landfilled and biogas will be vented through pipes; (2) 60% of paper, cardboard and plastics will be recycled and the rest will be landfilled, 40% of organic waste will be composted, and the rest landfilled. A 15-year period was selected as the solid waste disposal time (active operation time) of the hypothetical landfill in both scenarios. Waste composition is assumed to remain constant during the landfill operation period 2016–2031. Due to the small size of the required landfill in each island and the technological limitations present in isolated locations, landfill gas recovery was not considered a viable option (Bo-Feng et al. 2014; Mohee et al. 2015).

Landfill methane gas generation was estimated according to the first-order Scholl-Canyon model (IPCC 2006). This model is extensively used for estimation of landfill gas generation and has provided better results than other methods such as zero-order methods (Thompson et al. 2009; Kumar and Sharma 2014). Scholl-Canyon model is based on a first-order decay of methane generation and can be expressed by the following equation:

$$Q = M k L_0 e^{-kt} \quad (8.1)$$

where Q stands for methane generation (ton year^{-1}), M is the solid waste deposited in the landfill (ton year^{-1}), k is the first-order methane decay rate (year^{-1}), L_0 is the methane generation potential (ton$_{CH4}$ ton$_{waste}^{-1}$), and t is the time after solid waste disposal (year). A modified Scholl-Canyon model, the LandGEM model, is recommended by the US Environmental Protection Agency (USEPA 2005) and divides the amount of waste M by 10 (Eq. 8.2):

$$Q = \frac{M}{10} k L_0 e^{-kt} \quad (8.2)$$

Thompson et al. 2009 have shown that the mean absolute error of the model compared to methane recovery rates of 35 different landfills is minimized when the divisor ranges between 1.5 and 2.3 (a divisor equal to 10 resulted in much larger absolute errors). For the present study, a modified Scholl-Canyon model with a divisor $n = 2$ was selected (see Eq. 8.3):

$$Q = \frac{M}{n} k L_0 e^{-kt} \qquad (8.3)$$

The total methane generation was determined as the summation of the generated methane arising from disposed waste in the landfill every half year (divisor $n = 2$), that is:

$$Q_{total} = k L_0 \sum_{i=1}^{N} \frac{M}{2} e^{-kt_i} \qquad (8.4)$$

where N is the number of periods (half year periods) required for landfill stabilization. N equals 100 for a stabilization time of 50 years, which was selected for this study (McBean et al. 1994). The methane generation potential ($ton_{CH4}\ ton_{waste}^{-1}$), L_0, is the amount of methane that is generated by the anaerobic degradation of a certain amount of solid waste. This was determined for rapidly biodegradable waste (organics and paper and cardboard) and slowly biodegradable waste (residual waste) from the element composition of their biodegradable volatile solids. A mass balance equation was used for determining the amount of methane (Eq. 8.5):

$$C_a H_b O_c N + \left(\frac{4a-b-2c+3}{4}\right) H_2 O$$
$$\rightarrow \left(\frac{4a+b-2c-3}{8}\right) CH_4 + \left(\frac{4a-b+2c+3}{8}\right) CO_2 + NH_3 \qquad (8.5)$$

A methane decay rate k (year^{-1}) for rapidly and slowly biodegradable waste corresponding to half-life times, $t_{1/2}$, of 5 and 20 years, respectively, was selected (Scharff and Jacobs 2006):

$$k = \frac{\ln 2}{t_{1/2}} \qquad (8.6)$$

Scenario 1: Landfill

In this scenario, organics and paper and cardboard are assumed to biodegrade rapidly. Plastics and glass are considered as nonbiodegradable, whereas the rest is considered to biodegrade slowly (Tchobanoglous et al. 1998). Humidity (%H), volatile solids (%VS), biodegradable volatile solids (%BVS), and element composition of the rapidly and slowly biodegradable components for Santa Cruz, San Cristobal, and Isabela are presented in Table 8.3. Humidity varied between 50% and 65% (percent by weight as collected) for rapidly and slowly biodegradable components in all islands, whereas volatile solid ranged between 29% and 41%. The percentage of biodegradable

Table 8.3 Solid waste characteristics (percent by weight as collected)

	%H	%VS	%BVS	L_0 $ton_{CH4}\ ton_{BVS}^{-1}$	k $year^{-1}$	Element composition of BVS
Santa Cruz						
Rapidly biodegradable fraction	61	37	26	5.71×10^{-2}	0.139	$C_{27.0}H_{43.6}O_{17.0}N$
Slowly biodegradable fraction	50	41	15	4.73×10^{-2}	0.035	$C_{24.5}H_{39.3}O_{14.9}N$
San Cristobal						
Rapidly biodegradable fraction	65	34	24	8.22×10^{-2}	0.139	$C_{24.5}H_{39.4}O_{15.0}N$
Slowly biodegradable fraction	54	36	13	4.73×10^{-2}	0.035	$C_{22.9}H_{36.7}O_{13.7}N$
Isabela						
Rapidly biodegradable fraction	69	29	22	5.71×10^{-2}	0.139	$C_{22.4}H_{35.9}O_{13.3}N$
Slowly biodegradable fraction	65	33	12	4.73×10^{-2}	0.035	$C_{21.7}H_{34.8}O_{12.8}N$

volatile solids varied between 12% and 26%. L_0, the methane generation potential, ranged between 4.73×10^{-2} and 8.22×10^{-2} $ton_{CH4}\ ton_{BVS}^{-1}$; this is equivalent to 115 to 66 $m^3_{CH4}\ ton_{BVS}^{-1}$ (1 atm and 273 K), which is in the expected range reported for organic wastes (Kumar and Sharma 2014; Thompson et al. 2009). The first-order methane decay rate, k, used for this study was 0.139 $year^{-1}$ for rapidly biodegrading waste and 0.035 $year^{-1}$ for slowly biodegrading waste. This is consistent with reported values of 0.187–0.231 $years^{-1}$ for rapidly biodegradable waste such as food and garden waste and 0.030 $years^{-1}$ for slowly biodegrading waste materials (Scharff and Jacobs 2006; Kumar and Sharma 2014).

Figure 8.3 shows landfill methane production in San Cristobal in scenario, 1 (Scholl-Canyon model, $n = 2$). Dotted lines stand for methane generation of rapidly biodegrading wastes deposited every 0.5 years (rapid methane generation), which shows a peak in the first year and a first-order methane generation decay afterward. The total rapid methane generation is obtained by adding up the generation of all wastes disposed in the 15 years of landfill active operation. Figure 8.4 depicts methane generation for rapidly and slowly biodegrading waste (slow methane generation) in San Cristobal for a period of 50 years (landfill stabilization time). The first 15 years correspond to the landfill active operation period and the following 35 years are the landfills after-closure period. Total and rapid methane generation peaks after 16 years, whereas slow methane generation peaks after 18 years. Rapid methane generation is 330% larger than slow methane generation at their peaks. Nevertheless after 16 years there is a steep decrease in rapid methane generation rate, which in year 30 equals slow biodegradable methane generation. In the following 20 years, most methane is generated by slow biodegradable organic matter. Figure 8.5 compares total methane generation (rapidly and slow methane generation) in the three islands. Methane

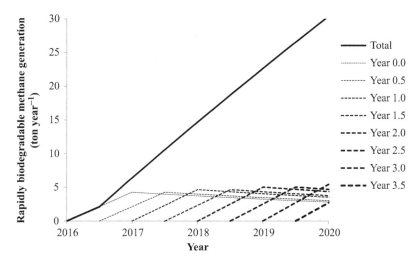

Fig. 8.3 San Cristobal rapid biodegradable methane generation Scholl-Canyon model ($n = 2$). Scenario 1

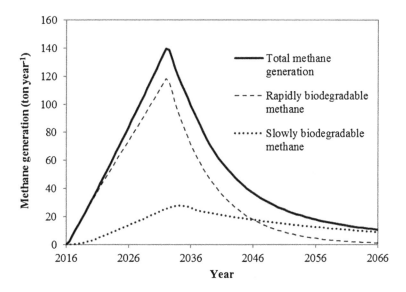

Fig. 8.4 San Cristobal slow methane generation—Scholl-Canyon model ($n = 2$). Scenario

generation in all islands increases rapidly and peaks 16 years after waste deposition. The highest methane generation takes place in Santa Cruz, which at its peak is 2.4 and 4 times larger than in San Cristobal and Isabela, respectively. Accumulated methane generation is depicted in Fig. 8.6. In Santa Cruz, San Cristobal, and Isabela, 11,526, 5131, and 2588 tons of cumulative methane are correspondingly generated within the 50-year landfill stabilization time.

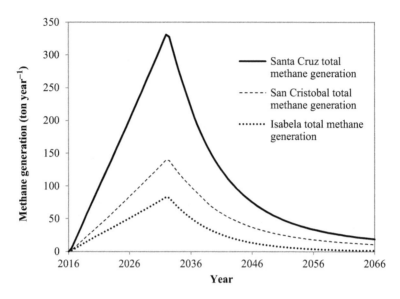

Fig. 8.5 Galapagos total landfill methane generation—Scholl-Canyon model ($n = 2$). Scenario 1. Fifty-year landfill stabilization time

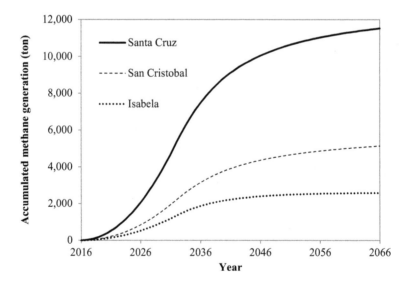

Fig. 8.6 Galapagos accumulated landfill methane generation—Scholl-Canyon model ($n = 2$). Scenario 1. Fifty-year landfill stabilization time

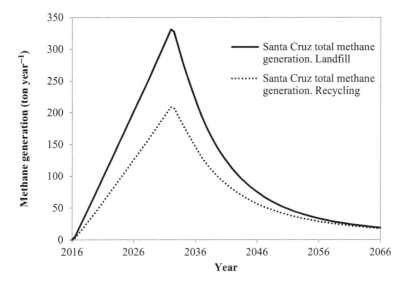

Fig. 8.7 Galapagos total landfill methane generation—Scholl-Canyon model ($n = 2$). Scenario 1 (landfill) and scenario 2 (recycling). Fifty-year landfill stabilization time

Scenario 2: Recycling

In scenario 2, 40% of organic matter is composted, 60% of recyclable waste (paper, cardboard and plastics) is separated and transported to the mainland, and the rest is disposed of. Therefore, 60% of organics, 40% of paper, cardboard, and the residual wastes are landfilled and degrade anaerobically to produce biogas. For this scenario, the accumulated solid waste generation quantities for the 15-year period were 85,770 tons in Santa Cruz, 43,205 tons in San Cristobal, and 11,242 tons in Isabela. Methane generation is shown in Fig. 8.7 for Santa Cruz scenario 2 (recycling), and it is compared to scenario 1 (landfill). Methane generation peaks in both scenarios at year 16. At the peak, methane generation for scenario 1 amounts to 331 ton year^{-1}, which is 59% larger than methane generation in scenario 2. Figure 8.8 shows the accumulated landfill methane generation for Galapagos (total of the three islands). After 16 years, the accumulated methane generation for scenario 2 is 38% lower than in scenario 1. Nevertheless, this percentage decreases with time, so that at year 50 the accumulated generation is 33% smaller for the recycling scenario. Table 8.4 shows the accumulated landfill methane generation for the three islands 50 years after initial waste disposal. Rapid methane generation for scenario 1 equals 78% of total methane generation, whereas for scenario 2 this percentage diminishes to 67%. Furthermore, rapid methane generation in scenario 2 equals 58% of the rapid methane generation in scenario 1. Santa Cruz and San Cristobal generate 87% of all landfill methane in both scenarios. In contrast Isabela generates only 13%.

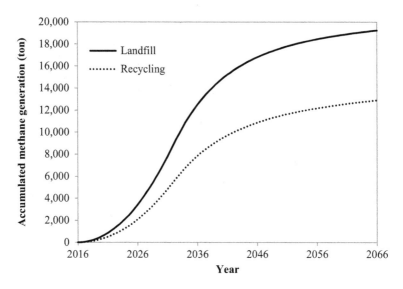

Fig. 8.8 Galapagos accumulated methane generation for scenario 1 (landfill) and scenario 2 (recycling)

Table 8.4 Landfill methane generation scenarios 2016–2066*

		Santa Cruz	San Cristobal	Isabela
Scenario 1: landfill	Rapid methane generation (ton)	8822	3670	2517
	Slow methane generation (ton)	2705	1498	71
	Total methane generation (ton)	11,527	5168	2588
Scenario 2: recycling	Rapid methane generation (ton)	5036	2141	1502
	Slow methane generation (ton)	2705	1498	71
	Total methane generation (ton)	7741	3624	1573

*Fifty-year landfill stabilization time frame

Conclusions

A population increase of 30% to 40% in Galapagos Islands is expected for the next 15 years, while a larger raise in solid waste generation is forecasted ranging from 160% to 180%. For scenario 1, Santa Cruz generates 123.5×10^3 ton of waste; San Cristobal generates 58.4×10^3 ton; Isabela generates 18.9×10^3 ton. By the use of recycling (scenario 2), a reduction in 30% of waste disposed to the landfill may be achieved.

According to a modified Scholl-Canyon model ($n = 2$), landfill methane generation rate in Galapagos peaks after 16 years of waste disposal and amounts 331 ton year^{-1} for scenario 1 (landfill). In scenario 2 (recycling), a 37% decrease in methane generation rate (at the peak) can be obtained. The rapidly biodegradable fraction generates 78% of the total accumulated methane generation by the landfill option, while recycling diminishes this to 67%. In both scenarios, Santa Cruz and San Cristobal generate 87% of accumulated landfill methane.

Due to the small size of the landfills in Galapagos Islands involved in the 15-year waste disposal period and considering the technological restrictions in isolated environments, landfill gas collection and use is not considered a viable. In order to reduce GHG emission from landfills in Galapagos, enhanced use of organic waste, especially in Santa Cruz and San Cristobal, is a better option.

Acknowledgments This work has been developed with the support of the University of Calgary and the Galapagos Institute of Arts and Sciences (GAIAS) of the Universidad San Francisco de Quito in the framework of the "World Energy Summit in the Galapagos, Ecuador, July 20–24, 2014."

References

Baine M, Howard M, Kerr S, Edgard G, Toral V (2007) Coastal and marine resource management in the Galapagos Islands and the Archipelago of San Andres: Issues, problems and opportunities. Ocean Coast Manag 50:148–173

Bo-Feng C, Jian-Guo L, Qing-Xian G, Xiao-Qin N, Dong C, Lan-Cui L, Ying Z, Zhan-Sheng Z (2014) Estimation of methane emissions from municipal solid waste landfills in china based on point emission sources. Adv Clim Chang Res 5(2):81–91

Chen M, Ruijs A, Wesseler J (2005) Solid waste management on small islands: the case of Green Island, Taiwan. Resour Conserv Recycl 45:31–47

Czepiel PM, Shorter JH, Mosher B, Allwine E, McManus JB, Harriss RC, Kolb CE, Lamb BK (2003) The influence of atmospheric pressure on landfill methane emissions. Waste Manag 23(7):593–598

Dirección del Parque Nacional Galápagos (2014) Plan de Manejo de las Áreas Protegidas de Galápagos para el Buen Vivir. Puerto Ayora, Galápagos, Ecuador

INEC (2010) National Institute for Statistics and Census of Ecuador Resultados del censo 2010 de población y civienda en el Ecuador. Fascículo provincial Galápagos. www.inec.gob.ec

Intergovernmental Panel on Climate Change (IPCC) (2006) IPCC Guidelines for National Greenhouse Gas Inventories, vol 5 Waste http://www.ipcc-nggip.iges.or.jp/public/2006gl/vol5.html (retrieved 25.02.2016)

Kumar A, Sharma M (2014) Estimation of GHG emission and energy recovery potential from MSW landfill sites. Sustain Energy Technol Assess 5:50–61

McBean E, Rovers F, Farquhar G (1994) Solid waste landfill engineering and design. Prentice Hall, Englewood Cliffs, NJ

Ministry of Environment of Ecuador (MAE) (2016) http://www.ambiente.gob.ec (retrieved 25.02.2016)

Mohee R, Mauthoor S, Bundhoo Z, Somaroo G, Soobhany N, Gunasee S (2015) Current status of solid waste management in small island developing states: a review. Waste Manag 43:539–549

Santamarta J, Rodríguez-Martín J, Arraiza M, López J (2014) Waste problem and management in insular and isolated systems. Case Study in the Canary Islands (Spain). 2014 International Conference on Environment Systems Science and Engineering. IERI Procedia 9:162–167

Scharff H, Jacobs J (2006) Applying guidance for methane emission estimation for landfills. Waste Manag 26:417–429

Tchobanoglous G, Theisen H, Vigil S (1998) Integrated solid waste management. McGraw-Hill, New York

Thompson S, Sawyer J, Bonam R, Valdivia J (2009) Building a better methane generation model: validating models with methane recovery rates from 35 Canadian landfills. Waste Manag 29:2085–2091

USEPA (2005) First-order kinetic gas generation model parameter for wet landfills. EPA-600/R-05/072

Chapter 9
Biodigesters as a Community-Based Sustainable Energy Solution

Elizabeth Romo-Rábago, Irene M. Herremans, and Patrick Hettiaratchi

Introduction

The development of alternative energies is becoming more important as the use of fossil fuels are becoming recognized as a link to climate change. Alternative energies can also contribute to poverty abatement. According to Takada and Charles (2006), biomass (energy produced from organic sources such as wood, waste, garbage, or animal matter) is recognized as the prime source of energy for the poor. A biodigester is the technology used to recover methane and other by-products from animal waste through anaerobic digestion (Fig. 9.1).

This process stabilizes the organic matter, reduces pathogens and odors, and reduces the total solids and sludge quantities by converting part of the volatile solids to biogas (Burke 2001; Rowse 2011). It is estimated that by 2050, sustainable biofuel and biomass production could add 100 EJ (ExaJoule) to the global energy supply with few or no net CO_2e emissions.

However, there are many challenges for developing biomass as an effective sustainable source of energy. One of them is implementing effective technology best suited to the end user and the location.

E. Romo-Rábago (✉)
The Social License Consortium, Calgary, AB, Canada
e-mail: biologaliz@gmail.com

I.M. Herremans
Haskayne School of Business, University of Calgary, Calgary, AB, Canada
e-mail: irene.herremans@haskayne.ucalgary.ca

P. Hettiaratchi
Department of Civil Engineering, Schulich School of Engineering, University of Calgary, Calgary, AB, Canada
e-mail: jhettiar@ucalgary.ca

© Springer International Publishing AG 2018
M.-E. Tyler (ed.), *Sustainable Energy Mix in Fragile Environments*, Social and Ecological Interactions in the Galapagos Islands,
https://doi.org/10.1007/978-3-319-69399-6_9

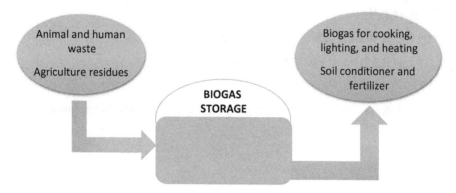

Fig. 9.1 Diagram of a typical low-cost biodigester. Reproduced from Romo-Rabago (2014)

It is well known that the lowest-income populations spend a substantial share of their income and/or time on low-efficient energy (Daisy and Kamaraj 2011). About 79% of the lowest-income households use wood almost exclusively for cooking, which confirms the dependency on biomass for basic survival activities. Wood provides 15% efficiency for cooking, while biogas provides 60% efficiency. Biogas can be generated by natural occurring degradation of human and animal waste. Anaerobic digestion appears to be the most suitable alternative for waste treatment and biogas generation (Takada and Charles 2006; Rowse 2011; Daisy and Kamaraj 2011).

The use of biodigesters has a long history dating back 2000–3000 years ago to ancient China. In 1859, the first digestion plant was built at a leper colony in Bombay, India. However, the high cost and the time-consuming nature of constructing a biodigester was an important factor in limiting its distribution.

More recently, during the 1970s, biodigesters gained popularity in developed countries, but the high cost and complex technology used for biodigesters prevented developing countries from accessing them. It was not until the late 1980s that low-cost biodigesters were developed.

This technology is currently growing in popularity and offers developing countries a sustainable source of energy that could improve quality of life (Harris 2012).

Some important characteristics of low-cost biodigesters are that they do not require a heating system or mixing mechanisms such as those used in largescale facilities which however, the costs for building a biodigester. In addition, a low-cost biodigester can provide a community with important benefits (Rowse 2011; Marti Herrero 2012) including:

1. Energy generated through methane production can be used for cooking and heating or electricity generation.
2. An odorless slurry is produced that can be used as a fertilizer for improving the crop productivity because it contains nitrogen and phosphorous.

3. Improved health benefits because biogas do not release toxic gases.
4. Women are empowered because in most developing countries, women and girls are responsible for cooking and a key element in the operation of the biodigester.
5. Animal manure can be disposed of safely; reducing smells, flies, and the spread of diseases.
6. Using fewer trees for cooking will reduce GHG emissions and provide environmental benefits.
7. Workload is reduced because the time needed for loading the biodigester with fresh manure is less than that required for collecting wood fire and is cheaper than buying fuel.
8. It is a sustainable technology that it is simple and uses local materials, reducing operational and maintenance costs.

However, to maximize these benefits, it is important that before installing a low-cost biodigester, local conditions are well understood. Factors such as climate, type of waste being digested, its concentration, and presence of toxic metals as well as socioeconomic and environmental conditions are very important criteria for selecting the best and most suitable biodigester (Pulamte and Abrol 2003).

Jiudao Yakou, China

The Jiudao Yakou village is located in southwestern China in Yunnan province, 800 km from Kunming city, its capital. This province is part of China's western provinces in which most of the poor people of China live in widely dispersed rural communities that have little infrastructure and low levels of access to modern energy services. At present, there are around 4.6 million households, mostly in remote areas far away from power grids, and lacking access to electricity.

Although the western provinces comprise China's poorest areas, they are also endowed with the country's richest renewable energy resources, including wind, hydro, solar, and biomass. Currently, biomass is satisfying most of rural China's energy demands, mainly in the traditional forms of agricultural wastes and forestry residues (Global Network on Energy for Sustainable Development [GNESD] 2002).

Yunnan is China's most biologically and culturally diverse province. Although, Yunnan Province has less than 4% of the land of China, it contains about half of China's birds and mammals. The province has snow-capped mountains and tropical environments that support a variety of species. One of the ethnic groups is the Hani, who live in the mountainous area and have a long tradition of rice terrace agriculture (Liang 2011). However, Yunnan, similar to other provinces in China, is having problems due to fast economic development and urbanization. Since the 1990s, the ecological footprint of Yunnan changed from a surplus to a deficit which has increased rapidly. Intensive wood collection for fuel and building supplies, loss of habitat, and increases in energy consumption and animal food have been some of the unsustainable practices that have occurred (Ying et al. 2009).

Currently, the Jiudao Yakou village is struggling with three main issues. Water is the first and most important issue for the residents. Since this village does not have running water, their water sources come primarily from rainwater capture devices installed throughout the village. This water is used for cooking and watering corn crops through a cistern that each family at the village has access to. Attempts to hand drill for water have are limited because of the soil characteristics (limestone formation). The second issue is energy. Although the residents are connected to the electric grid, the service is uncertain because of weather conditions. Electricity transmission poles are prone to falling down, requiring weeks to fix and is not a reliable source of energy. Villagers spend about half of the day collecting firewood for cooking. There is also a need for a waste management plan to handle human and animal wastes because of a primitive latrines, and untreated human and animal wastes contain high concentrations of viruses and bacteria. Health risks increase when people are directly exposed to biological waste or when water supplies become contaminated (Eco Village 2012).

The installation of a low-cost biodigester offers a sustainable and affordable opportunity for treating and recycling human and animal waste, taking advantage of the organic content to produce fuel in the form of biogas. To achieve these benefits, a low-cost biodigester project has to incorporate social and biological systems characteristic of the village and an integral part of the technology transfer process.

San Cristobal, Galapagos Islands

The Galapagos Islands are a World Heritage Site, surrounded by an extensive marine reserve which is home to some of the world's most unique ecosystems. Some of the islands are inhabited. By 2006, three urban districts held 83% of the total population, with San Cristobal accounting for 36% (9,045) of Galapagos inhabitants. San Cristobal is the second most inhabited island after Santa Cruz, and a recent census estimated the population of Galapagos Islands at about 25,000 people. Of this total population, 17% live in the rural areas of the island (GNPS et al. 2013). The land area of San Cristobal is approximately 55,709 hectares, of which 80% is protected. The urban land area of San Cristobal is approximately 733.6 hectares, and is close to the highest point on the island (Teran 2008).

San Cristobal's environment is under stress and ecologically sensitive. Therefore, waste management is a critical issue. The current open dump on the island has several environmental problems, such as uncontrolled reproduction of cats, dogs, seagulls, and flies, bad odors, and uncontrolled leachate generation. Fossil fuels used for cooking come from the mainland, as people are dependent on outside energy sources.

Since the implementation of San Cristobal's solid waste management plan, there have been some issues regarding the operation of the composter facility and sanitary landfill. Poor planning in determining the size of the landfill has resulted in reaching its capacity. It can no longer manage non-recyclable waste after 8 years of operation (Peñafiel 2015; Zapata 2012). Currently, the solid waste facility and landfill are located 10 min away from Puerto Baquerizo on the borders of San Cristobal's urban area and 4.5 km toward El Progreso, the agricultural area of the island. The area where the facility and sanitary landfill exist is a total of 5 hectares. It is estimated that approximately 92% of the waste generated in San Cristobal comes from the urban and rural sectors, while 8% comes from the tourism and commercial sectors (Government of San Cristobal 2012).

Technology Transfer

Technology transfer is an interactive process between technology specialists and final users. This process is key for successful implementation of a technology. It is complicated because there is a gap between technology development and its application and weak interaction among scientists, external experts, and users (Pulamte and Abrol 2003). For example, in most biodigester developments, designers and administrators pay sufficient attention to the construction process; however, follow-up management, operations, daily maintenance, and repair work have not been well designed or carried out, which has resulted in inadequate technical services and support. Often, the social and cultural dimensions of technology transfer are ignored. This has resulted in the paradoxical situation of progress in biodigester construction, but households unable to actually benefit from it (Pulamte and Abrol 2003). For a biodigester to be a sustainable project, a holistic approach that considers the biosocial system, needs to be used. As Ehrenfeld (2008) suggests, if we want something to change, we needs to look at the structure that creates action.

In this case, to successfully implement a low-cost biodigester in the Jiudao Yakou village or the San Cristobal contexts, it is necessary to consider the culture of the residents and identify what is driving unsustainable practices related to water, energy, and waste management. The installation of a low-cost biodigester in specific contexts attempts to creates a more efficient waste management system with biogas generation for heat and cooking. Biodigesters have some risks and stresses such as biogas flammability, waste availability, inefficient management, maintenance, and operation. As well, it also requires a change in behavior to adjust to the new way of disposing of waste. Using Kofinas and Chapin's (2009) "adaptive capacity" framework (Fig. 9.2), the immediate impacts of these risks could be the biodigester's poor management, thus causing insufficient waste for generating biogas for the village. However, effective instructions for the installer, operator, and end user about safe operation of the equipment and its maintenance can improve the biodigester's management and eliminate or reduce the danger of biogas flammability. To have enough waste to generate biogas, a variety of waste is recommended.

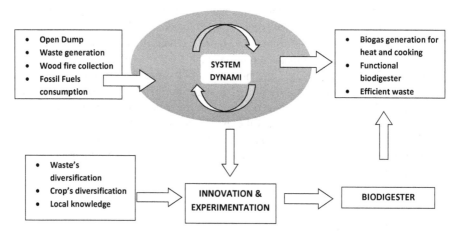

Fig. 9.2 Adaptive capacity framework. Reproduced from Romo-Rabago (2014)

Kofinas and Chapin (2009, p. 67) state that "Diversity provides the raw material or building blocks on which adaptation can act". Therefore, other sources of waste, such as human and animal waste, as well as by-products from crops, can be used to generate sufficient biogas. Finally, in regard to the new way of disposing of waste, social learning networks and local knowledge can be effective. One advantage the Chinese village has is that one of its residents has built a biodigester for his family and is ready to assist in the construction and operation of the biodigester. Moreover, residents are willing to change current beliefs and norms for something new that is socially desirable. Regarding the Galapagos context, there are a number of governmental and private sector organizations as well as academic institutions that are working hard to improve waste management on the island, including environmental education programs for the schools and the general public.

Environmental and Socioeconomic Benefits for the Jiudao Yakou Village

As mentioned earlier the Jiudao Yakou village is a leprosy community located in the Yunnan province, which is one of the world's biodiversity hotspots. However, Yunnan's natural resources are seriously under threat because its population highly depends upon the local ecosystem for its food and resources. Based on a feasibility study conducted in collaboration with the University of Calgary and the Eco Village of Hope Society, there are a number of benefits that a low-cost biodigester can bring to the village (Table 9.1). The installation of a low-cost biodigester at the Jiudao Yakou village has potential to provide villagers with a renewable energy source with multiple environmental and socioeconomic benefits, but it will also provide an

Table 9.1 Summary of potential biogas benefits at the Jiudao Yakou village, China

Biodigester's benefits at Jiudao Yakou village, China		
Environment	GHG reduction	71.35 t CO_2eq/year
	Deforestation reduction	8.3 ton/year = 138 trees
Economic	Fuel and fertilizer reduction	>5% savings
	Increase in crop yields	20–50%
Social	Reduction in time for firewood collection	>3 h/day
	Health improvement	CO, CO_2, SOx, H_2S, and PM below detection limits
Technical	Clean fuel for cooking	>60% fuel needs covered by biogas
	Clean fuel for lighting	

Reproduced from Romo-Rabago (2014)

efficient sanitary system that is relatively affordable and easy to operate and maintain. If successful, this project will be of interest throughout China, as an effective approach for sustainable energy development.

Environmental and Socioeconomic Benefits for San Cristobal

Similar to Jiudao Yakou village, San Cristobal has a fragile environment which is being impacted by human economic development and growth. Based on a feasibility study conducted in collaboration with the University of Calgary and the Universidad de San Francisco de Quito, a biodigester could bring multiple benefits to the island. Table 9.2 summarizes some possible benefits.

Conclusion

Biodigester technology transfer could be a sustainable solution for developing countries if biosocial conditions are well understood. In the case of the Jiudao Yakou village, as well as San Cristobal, the successful implementation of a low-cost biodigester depends on the adaptive capacity of the residents which involves innovation and experimentation (as in Fig. 9.2).

However, if it is successful, this technology will provide the residents with a sustainable source of energy that helps to alleviate the demand for fossil fuels and fuel wood and offers a more efficient waste management system. Therefore, the well-being of natural capital, as well as human capital will be improved because of avoidance of water, soil, and food crop contamination as a result of adequate human and animal waste management.

Table 9.2 Summary of potential biogas benefits at San Cristobal, Galapagos, Ecuador

Biodigester's benefits at San Cristobal, Galapagos, Ecuador (estimated)		
Environment	GHG reduction	1029 t CO_2eq/year
	Carbon sequestration	33,992m^3 CO_2/year
Economic	Diesel use reduction	25,367 L/year
	Liquid fertilizer generation	5700 L/day
Social	Job creation	>4 jobs created/1000 tons of waste recycled
	San Cristobal's landfill users	36% (9045) of Galapagos' inhabitants
Technical	Biogas generation	50,735m^3/year = 304,410 kWh
	Electricity generation	

Reproduced from Mesa-Lenis (2015)

References

Burke D (2001) Dairy waste anaerobic digestion handbook: options for recovering beneficial products from dairy manure. Retrieved from: www.makingenergy.com

Daisy A, Kamaraj S (2011) The impact and treatment of night soil in anaerobic digester: a review. Microb Biochem Technol 3(3):43–50

Ehrenfeld JR (2008) Sustainability by design. In: Culture change: locating the levers of transformation. Many Rivers Press, Langley, WS, pp 78–98

Global Network on Energy for Sustainable Development (2002) Poverty reduction: can renewable energy make a real contribution? Retrieved from: http://www.gnesd.org/downloadables/povertyreductionspm.pdf

GNPS, GCREG, CDF, GC (2013) Galapagos Report 2011–2012. Puerto Ayora, Galapagos, Ecuador. Accessed at: http://www.galapagospark.org/documentos/ciencia/GalapagosReport_2011-2012.pdf

Government of San Cristobal (2012) Gobierno Autónomo Descentralizado Municipal Del Canton San Cristobal, Plan de Desarrollo y Ordenamiento Territorial 2012–2016

Harris P (2012) Beginners guide to biogas. University of Adelaide, Australia. Retrieved from http://www.adelaide.edu.au/biogas/

Kofinas GP, Chapin FS (2009) Sustaining livelihoods and human well-being during social-ecological change. In: Chapin FS, Kofinas GP, Folke C (eds) Principles of ecosystem stewardship. Springer, Sweden, pp 55–75

Liang L (2011) Biodiversity in China's Yunnan province. Retrieved June 23, 2012 from United Nations University: http://unu.edu/publications/articles/biodiversity-in-chinas-yunnan-province.html

Marti Herrero J (2012) Low cost biodigesters to produce biogas and natural fertilizer from organic waste. Ideas latinamerica. Retrieved from http://www.ideassonline.org/public/pdf/BrochureBiodigestersENG.pdf

Mesa-Lenis GA (2015) Supervised by Herremans IM, Hettiaratchi P, Jayaraman M, Peñafiel R, Romo-Rabago BE, A Biodigester and Biocell Feasibility Study: A Sustainable Approach for Waste Management for the Island of San Cristobal, Galapagos, Ecuador. Master's thesis. University of Calgary, AB, Canada

Peñafiel R (2015) Email: information on San Cristobal. External Supervisor, Universidad de San Francisco de Quito

Pulamte L, Abrol D (2003) Technology transfer for rural development: managing R and D at CSIR. Econ Polit Wkly 38(31):3315–3318

Romo-Rabago BE (2014) Supervised by Herremans IM, Hettiaratchi P, Low cost biodigesters as a sustainable energy solution for developing countries: Jiudao Yakou village, China, a case study. Master's thesis, University of Calgary, AB, Canada

Rowse LE (2011) Design of small scale anaerobic digesters for application in rural developing countries. Unpublished master's thesis, University of South Florida, Tampa, FL

Takada M, Charles NA (2006) Energizing poverty reduction: a review of the energy poverty nexus in poverty reduction strategy papers. United Nations Development Programme, New York, NY

Teran A (2008) Gestión ambiental en la isla San Cristóbal: Islas Galápagos, Ecuador. Universidad Carlos III De Madrid: Escuela Politecnica Superior Departamento De Ciencia E Ingenieria De Materiales E Ingenieria Quimica

The Eco Village of Hope (2012) Retrieved from: http://www.ecovillageofhope.org/project1.html

Ying L, Daming HE, Buchanan S, Jiang L (2009) Ecological footprint dynamics of Yunnan, China. J Mt Sci 6:286–292

Zapata P (2012) Estudio De Impacto ambiental: Nuevo Sistema de Agua Potable para el Cantón San Cristobal, Provincia de Galápagos. Gobierno Municipal Del Canton San Cristobal. Accessed at: http://www.gobiernogalapagos.gob.ec/wp-content/uploads/downloads/2013/08/PDOT-San-Cristobal-2012.pdf

Chapter 10
Sustainable Energy Mix + Fragile Environments in Canada's Northern Coastal Zone: Is Technology Enough?

Mary-Ellen Tyler and Allan Ingelson

Northern Fragile Environments

Canada is officially 150 years old as a nation as of July 1, 2017. For the last 150 years, Canada's settlement history as a northern country rich in land and natural resources has clearly demonstrated the importance of natural resources, transportation, and energy in national patterns of settlement, economic development, and politics. Over the next 50 years, natural resources, transportation, and energy will continue to drive Canada's future evolution. However, in many ways, Canada is at a significant policy "crossroads" with respect to the social and cultural values, geopolitics, and global climate variability that will drive future natural resource and energy development.

Canada is bound on the south by the 49th parallel along its border with the continental United States of America (USA) and extends north of the 60th parallel to the Arctic Circle. Canada has the longest coastline in the world created by oceans on its western, eastern, and northern boundaries and the second largest landmass in the world (Statistics Canada 2014). However, its total population is approximately 36 million people, compared to the American state of California with an estimated population of 39.5 million people (United Nations 2017; World Population Review 2017). Canada's population is also highly urbanized with 80% located within 250 kilometers (km) of its southern border with the USA.

M.-E. Tyler (✉)
Faculty of Environmental Design, University of Calgary, Calgary, AB, Canada
e-mail: tyler@ucalgary.ca

A. Ingelson
Faculty of Law, Canadian Institute of Resources Law, University of Calgary, Calgary, AB, Canada
e-mail: allan.ingelson@ucalgary.ca

© Springer International Publishing AG 2018
M.-E. Tyler (ed.), *Sustainable Energy Mix in Fragile Environments*, Social and Ecological Interactions in the Galapagos Islands, https://doi.org/10.1007/978-3-319-69399-6_10

Historically, this settlement pattern reflects the relatively milder climate, longer growing season, and relatively accessible terrain of southern Canada that enabled construction of national transportation and energy infrastructure. Canada's population has historically clustered along the 49th parallel in the south. In contrast, the development of Canada's natural resources (forestry, mining, and energy) is geographically in the "near north" south of the 60th parallel, which encompasses the northern extent of Canada's ten provinces and the southern extent of Canada's federal territories – Yukon, Northwest Territories, and the eastern Arctic's self-governing Inuit territory of Nunavut.

Cold environments are "fragile" environments because long winters, short summers, low rates of available precipitation, and extremely low seasonal temperatures and light all contribute to limiting biogeochemical processes and ecosystem productivity. Canada's maritime Arctic coastal zone is a remote fragile area of approximately two million square km. It encompasses approximately 36,000 islands with many ecologically sensitive areas vulnerable to development and shipping impacts. This coastal zone is home to less than 30,000 people, who are primarily indigenous. It is an important habitat for marine mammals, seabirds, polar bears, and caribou. Due to extreme climatic and geographic conditions, there is almost no conventional transportation infrastructure, and air access is infrequent and expensive. The Canadian government's sea ice observations over the past 30 years show high year-to-year variability in sea ice coverage and indicate negative trends in ice coverage for both eastern and western maritime coastal regions. This sea ice loss could enable increased destination-specific seasonal Arctic shipping activity in the future (AMSA 2009). There is significant uncertainty associated with climatic projections, but AMSA scenarios suggest that within the next 10 years, resource development, oil and gas, ecotourism, fishing, and global shipping pressures will result in increased dry bulk, liquid bulk, supply/resupply, cruise, and container shipping seasonal activity. Year-round sea ice conditions, uncertainty, and risks will control the costs and feasibility of commercial shipping activity. "With the exception of nuclear ice breakers, very few ships have been built that could safely carry out year round commercial navigation in the Canadian Arctic" (AMSA 2009, p. 112).

Oil and Gas Reserves in Northern Fragile Environments

While the nineteenth and twentieth centuries in Canada's history have been shaped by the natural resources, transportation, energy, and development policies of the "south" and "near north," the twenty-first century will likely be shaped by the northern coastal zone of the "far north." Specifically, Bird et al. (2008) estimated the Arctic holds roughly 618 billion barrels of oil (BBO). An estimated 2.5 BBO previously identified in the Northwest Canadian Interior Basins have not yet been put into production. The Artic contains three times as much natural gas as oil on an energy-equivalent basis (Nong 2011). Imperial Oil based in Calgary, Alberta, has already expressed interest in the development of ultra-deep offshore oil drilling on

Canadian leases in the Beaufort Sea that could produce more than a billion barrels of oil at an estimated development cost of at least ten billion dollars (Dawson 2014).

There are similarities between remote fragile environments whether they are cold climate, hot arid climate, or small island developing states (SIDS) related to sustainable energy mix and fragile environment issues. Muir (2013) has identified three similarities:

- The need to adapt to climate variability including changing precipitation patterns and higher temperatures;
- The need for investment, technology transfer, and capacity development to support renewable energy technologies;
- A dependency on hydrocarbons to generate electricity.

Although manifested differently in significantly different situations, these three issues are exacerbated by problems of access and accessibility associated with geographic isolation in rural and remote contexts that contribute to high prices for domestic energy, transportation, and basic goods and services.

In a Canadian context, Natural Resources Canada (2011, p. 4) identified 292 remote communities defined as not connected to the North American electrical grid or to a piped natural gas network. Based on Statistics Canada 2006 Census reports, these "off-grid" northern and northern coast communities have a total population of 194,281 of which 65% are Aboriginal (First Nations, Metis, and Inuit peoples). All of these communities are located within an area "extending from over 20 degrees of latitude and 90 degrees of longitude and from artic dry climate to sea-coast humid forestry climate and from mountainous to plain regions"(Natural Resources Canada 2011, p. 38) which represents portions of seven Canadian provinces and three territories. A total of 251 or 86% of these communities have their own fossil fuel power plants producing 453.3 MW (Natural Resources Canada 2011, p. 6) which are subsidized to varying rates and levels by federal, provincial, and territorial governments.

All of these communities have a high degree of dependency on imported fuel and its associated high costs. The cost of producing off-grid energy is estimated to be ten times higher than North American grid costs and the subsidized consumer rates estimated at three times more than rates paid by on-grid consumers. Muir (Natural Resources Canada 2011, p. 11) estimates approximately 75% of Canada's northern fuel consumption is imported refined hydrocarbons. Between 2001 and 2012, the National Research Council of Canada reports that the wholesale price of diesel fuel increased by 132%.

The most significant feature of Canada's northern and remote environment energy status is the conclusion made by the off-grid working group of the Renewable and Electrical Energy Division, Energy Policy Sector, Natural Resources Canada (2011, p. 38) which stated "…little has been done to integrate local resources into the energy mix of these communities." Also of significance are the barriers that have been identified as preventing a more sustainable energy mix. Specifically, the Clean Tech Community Gateway (CTCG) report identified four primary barriers preventing remote Aboriginal communities in northern coastal British Columbia (including

Haida Gwaii) from benefiting from renewable energy: "limited financial capacity, limited ownership, limited human capacity, logistics in remote communities" (CTCG 2012, p. 2).

Technology is certainly implicit in any successful sustainable energy transition. However, both Canadian and international rural and remote energy research reports clearly identify the barriers, and lessons learned are not primarily technical. For example, an extensive international report on renewable energy for remote areas and islands (IEA-RETD 2012) identifies socioeconomic, institutional, financial, and environmental "lessons learned" from case studies in a diversity of locations and climatic conditions including island case studies from Canada's northern coastal zone (Ramea Island), Scotland (Isle of Eigg), and Galapagos (Floreana). The findings from these case studies identify risk perception and risk mitigation as critical factors in financing energy transition in remote environments. Specifically, risk is related to the cost of capital, insofar as a perception of high risk increases the cost of capital which in turn affects "the initial affordability, competitiveness, as well as the levelized costs of renewable energy" (de Jager and Rathmann 2008 in IEA-RETD 2012, p. 97). As a result, five types of risk were identified in the case studies and are summarized in Table 10.1. In the context of Canada's remote northern and northern coastal zone communities, high upfront costs for renewable energy projects related to the types of risk in Table 10.1 make it difficult to displace existing subsidized diesel systems.

Table 10.1 Types of risk for renewable energy financing in remote environments

Project risk
- Remote and fragile environment locations prone to delays related to distance and geographic location
- Extreme climates and remote locations exposed to seismic events and major seasonal weather events resulting in delays or uncertainty of logistics or availability of expertise
Construction risk
- Remote areas likely to have higher construction risks due to limited and uncertain availability of materials, labor, and transportation resulting in potentially significant cost overruns
Revenue risk
- Price-specific contracts for remote areas more difficult to secure
- Counterparties (the utility or business purchasing the power) in remote locations may be too small to be creditworthy or able to pay
- Weather conditions and other environmental factors may have disproportionate impacts in remote areas
Operational risk
- Reduced availability of trained on-site expertise and availability of spare parts can affect operational reliability and increase downtimes
Political risk
- An absence of credible institutions and policy or legal frameworks may increase perception of risk

Source: Adapted from IEA-RETD (2012, p. 101)

Closely associated with the importance of risk perception is the importance of social license and community acceptance which is a similar barrier to and critical factor in transforming energy mix in remote and fragile environments. This is particularly relevant in Canada's northern coastal zone where the population is predominantly of Aboriginal ancestry. The cultural values, world view, and governance in traditional Aboriginal lands are directly related to natural resource use and actively practiced in the north in large part because it is so remote from the urbanized populations to the south.

As the world's fifth largest oil and gas producer with the third largest proven oil and gas reserves in the world, energy development will continue to drive Canada's economy (Eyford 2013). For example, natural gas provides half of Canada's residential heating demand and half of the country's industrial activities (Canadian Energy Petroleum Association in Eyford 2013, p. 13). However, as Canada's historical natural gas export market to the USA rapidly diminishes, international demand for liquefied natural gas (LNG) "is expected to almost double by 2040" (Eyford 2013, p. 10). Therefore, if Canada is to build new natural gas export markets, pipelines and tidewater terminals will be needed primarily in Canada's northern coastal zone. The potential scale of LNG tidewater development as well as the future development of other known oil and gas reserves and mining resources in Canada's "far north" is massive to contemplate. Such development will directly affect, if not transform, Canada's northern marine and terrestrial environments as well as the current and future populations of the Aboriginal peoples living there.

Situational Energy Demand

Canada's remote northern coastal zone region is potentially one of the most rapidly changing environments on the planet in large part due to environmental stressors from changing climatic conditions. These changes are likely to drive up the cost and increase the uncertainty of hydrocarbon-dependent energy. In addition, other infrastructure needs including water, wastewater, and transportation are also likely to become more uncertain and expensive with changing environmental conditions. For example, the international Arctic Council's (2013) interim report on "Arctic Resilience" has identified extensive and expensive damage to infrastructure in both Alaska and Russia from climate change and mitigation measures. Specifically, the Arctic Council projects extra costs for infrastructure replacement due to climate change to 2030 as 30% for water and sewage, 25% for roads, 23% for airport runways, 8% for harbors, and 13% others (Arctic Council's 2013, p. 82).

Global socioeconomic demand for resources is also a major factor affecting the future of Canada's northern and remote communities. Specifically, Hausner et al. (2011) suggest increased world demand, together with favorable environmental conditions, and decreased production in other parts of the world (for geopolitical or other reasons) are sufficient to drive up oil prices to the point where the exploitation of Arctic resources on a large scale may be possible for the first time. If this were to occur, growth and development pressures would not only affect existing northern communities but would drive the construction of new towns and large worker "camps" similar to those currently operating in Alberta's oil sands. These new settlement patterns and their related infrastructure needs would be necessary to service the construction and operation phases of offshore drilling, tidewater ports, pipelines, and transportation infrastructure. All of this development would create new and increasing demands for a sustainable energy mix capable of supporting both remote domestic and industrial energy needs on a massive geographic scale.

Canada's northern coastal zone communities experience extremely cold seasonal temperatures and very few daylight hours. This translates into higher energy consumption during the winter months and a higher cost of living. The scale of geographic distance and connectivity factors and the current dependency on external fuel delivery by truck, air, or boat result in high costs for northern Canadians. Extremely cold seasonal temperatures and extreme weather conditions can also present challenging conditions for the performance and maintenance of renewable energy technologies such as wind turbines and solar panels.

This stands in stark contrast to the situation in Canada's provinces to the south. For example, the province of Ontario represents almost 40% of Canada's total population and contains much of the country's historical industrial heartland and its electricity demand. In 2010, the world's largest solar photovoltaic (PV) facility was operating in the City of Sarnia in Southern Ontario. In 2009 (Solar Energy 2014), Ontario adopted a provincial *Green Energy Act* (Ontario 2009) and a feed-in tariff (FIT) system for electricity rates from solar and wind sources. The intent of an FIT arrangement is to set a fixed price for energy from renewable sources and ensure the price paid to renewable electricity developers is not subject to market fluctuations.

However, recent legal challenges have arisen in Ontario specific to wind energy development. The provincial government is experiencing opposition from residents and dealing with lawsuits related to concerns about wind turbine development effects on human health and safety and wildlife. For example, in the case of *Hanna v Ontario*, people living in proximity to a proposed wind farm challenged whether the 550 meter setback distance from a residence was adequate to protect human health (ONSC 609 2011). Similarly, in 2014, an Ontario regulatory decision was challenged in court due to concerns about the impact of a wind project on "Blanding's turtle" (*Emydoidea blandingii*), a species found in the proposed project area and listed as threatened and protected under Ontario's *Endangered Species Act* (Ontario 2007).

Other Canadian provinces are also involved in renewable energy development. Canada's third largest province is British Columbia (BC) which is located on the west coast and has 13.7% of Canada's population. Similar to Ontario, BC has also taken policy and legislative initiatives to promote increased renewable energy development. In 2010, the BC provincial government adopted its provincial *Clean Energy Act* (BC Clean Energy Act) with provision for an accompanying FIT program. BC's legislation created a framework to establish renewable energy technologies as a basis for the province's energy future. Specifically, BC's two primary provincial energy policy goals are to establish a minimum of 93% of electricity from clean or renewable sources and to foster the development of innovative technologies to reduce provincial greenhouse gas emissions through renewable resource use. To this end, between 2008 and 2012, BC's *Innovative Clean Energy (ICE) Fund* has provided $75 million to 62 renewable energy projects to support "pre-commercial" energy technology development and deployment of new commercial technologies in solar, wind, geothermal, tidal, and bioenergy not currently in use (Government of Canada 2014). However, similar political directives, policy tools, and framework legislation have not been initiated by Canada's federal and territorial governments for northern remote communities.

Renewable Energy Sources

Canada's abundant hydroelectric energy resources account for the majority of electricity generation in Canada from sources in the provinces of Quebec, British Columbia, Manitoba, Newfoundland and Labrador, and Yukon Territory (National Energy Board 2013). In 2012, 62% percent of Canada's national electricity generation capacity was from renewables (National Energy Board 2013). Between 2008 and 2012, non-hydro renewables (wind, solar, and biomass) were the fastest-growing generating sources in Canada with an annual growth rate of 16% (National Energy Board 2013, p. 67). In "Canada's Energy Future 2013: Energy Supply and Demand Projections to 2035," the National Energy Board forecasts additional wind development will make the most significant contribution to non-hydro renewable electricity growth from 2013 to 2035 (National Energy Board 2013).

The potential for tidal energy development has been tested at the Fundy Ocean Research Center for Energy (FORCE) in the province of Nova Scotia since 2006. The Bay of Fundy is part of Nova Scotia, an island province on Canada's east or Atlantic Ocean coast. The Bay of Fundy has the highest tides in the world, and an estimated 100 billion tonnes of water moves in and out of the bay during every 12-h. tidal period. Despite complex seabed conditions, FORCE has been testing subsea turbines since 2009 with the goal of developing the world's largest in-stream tidal power infrastructure (Marine Renewables Canada 2013).

Fossil fuels (oil, gas, and coal) are the dominant electricity generation sources in Canada's two western prairie provinces of Alberta and Saskatchewan. Currently, coal and natural gas account for more than 80% of electricity generation in the province of Alberta. However, renewable energy sources are beginning to be recognized as having an increasing role to play. For example, a review of Alberta's electricity generation sources as of November 2013 shows the following renewable source contributions: hydroelectricity 6.4%, wind 7.9%, and biomass 3%. Currently in Alberta, 29 out of a total of 72 planned electricity-generating projects represent renewable energy sources (3000 MW). This initial phase of a transition in energy mix has primarily been a low-carbon response to climate change. Specifically, the province of Alberta adopted the "Climate Change Emissions Act" in 2003 with the accompanying creation of carbon dioxide emissions limits using a "Specified

Emitters Regulation" mechanism to enforce financial penalties on industrial CO_2 emitters including oil and coal companies.

Canada's energy sources have developed over the past 150 years to support its distinct historical patterns of population and agricultural and industrial activity in response to significantly different regional conditions. The challenge of Canada's "far north" and its northern coastal zone has yet to become the focus for Canada's national resource development interests and population growth. However, the growing provincial interest, investment, and expertise in renewable energy generation and energy mix transition in southern Canada are creating a large-scale body of knowledge and experience that can be adapted to northern regions.

Factors Affecting Energy Mix Planning and Implementation

Energy forecasting by Canada's federal energy regulator, the National Energy Board (NEB), indicates a significant reduction in coal-generated electricity production between 2013 and 2035 and a significant commensurate increase in generation from natural gas and wind. The experience and expertise gained by both public and private sector actors in provincial renewable energy development represent a major opportunity for the knowledge and technology transfer process that will be necessary for large-scale northern energy transition. However, the nature of the organizational, policy, and institutional framework within which future northern energy planning, management, and development might occur and who might be involved needs to be considered. To date, Canada's federal approach to energy policy has been based on three principles (Natural Resources Canada 2014):

- Competitive markets are assumed to be the most efficient mechanism for regulating supply, demand, and exports and meeting national energy needs.
- Provincial jurisdiction is essential and provincial governments are responsible for managing most of the country's energy resources.
- Federal intervention through regulations/standards and economic incentives/disincentives will only be used as necessary to ensure that specific public health, safety, and environment objectives are achieved.

Key to understanding energy management planning and regulation in Canada is the powers and roles of the provinces and the federal government as defined in the Constitution Act (1867, 1982). In the case of energy, both levels of government are responsible to plan for and regulate energy development. Section 92A of the Constitution Act, 1982, confers on each provincial government the exclusive authority to make laws in relation to the "development, conservation and management of non-renewable natural resources and forestry resources in the province … and … [the] development, conservation and management of sites and facilities in the province for the generation and production of electrical energy" (Benidickson 2013, p. 38). The courts have also recognized provincial legislative authority to regulate most of the energy development in provinces based on their ownership of oil, natural gas, coal, other minerals, and other natural resources such as water and forests on provincial lands as provided under the constitution (Constitution Act 1867, Section 109). In southern Canada, most of the land base is under provincial jurisdiction, and most of the land use planning for energy development and regulation of energy projects is undertaken by and is the responsibility of provincial governments. However, the federal government has the power to plan for and regulate energy development on federal land which includes the territories in northern Canada that have not attained provincial status. In addition to the territories, the federal government regulates energy development on other federal lands within provincial boundaries such as national parks, military bases, and Indian reserve lands.

The federal government is responsible for Canada's offshore and coastal interests along Canada's continental shelf. The federal government is also responsible for protection of Canada's water and fishery resources (including seacoast and inland fisheries), navigation and shipping, trade and commerce, and the implementation of treaties between Canada and other countries. Environment and Climate Change Canada is the federal department responsible for (1) coordinating federal policies to preserve and enhance Canada's environment, (2) conserving national natural resources including water, and (3) regulating GHG emissions (Environment Canada 2012). A second federal department, Natural Resources Canada (NRC), is similarly responsible for taking the lead on the development and implementation of federal energy policies as they relate to natural resource development.

The federal government transferred control of public lands, resources, and water to the territorial government of the Northwest Territories via the Northwest Territories Devolution Agreement and Devolution Act (2014). Under this Agreement, oil and gas rights are deemed as "onshore" and therefore can be administered by the territorial government. However, administration of oil and gas rights in Nunavut and the Arctic Offshore remains under federal government authority.

The primary federal legislation related to the planning and approval of new energy projects under federal jurisdiction is the Canadian Environmental Assessment Act (CEAA) which came into force July 6, 2012, and replaced the previous 1992 CEAA legislation. The revised 2012 legislation was a response to the federal government's concerns about delays in the approval of major energy infrastructure projects. As a result, the previous federal practice of broad public consultation was replaced with consultation limited to project proponents and industry associations and consistent with the objectives of the federal government's Jobs, Growth, and Long-term Prosperity Act (Government of Canada 2012b).

- *To promote investment in the energy sectors by increasing efficiency in the federal EA process*
- *To promote increased cooperation and coordinated EAs between the federal and provincial governments*
- *To ensure EAs are completed in a timely manner*

Under the revised CEAA 2012 (Section 19), the factors which will be considered in an EA process applicable to major energy projects include the following:

- Purpose of the proposed energy project;
- Environmental effects (including malfunctions and accidents scenarios) and their significance;
- Public comments;
- Mitigation measures
- Requirements for the follow-up program;
- Alternative means of carrying out the energy project; and
- Results of relevant studies and any other relevant matters.

The revised CEAA 2012 includes provision for the jurisdictional relationship between federal and provincial governments. Under Section 32, a provincial EA process may be substituted for a federal EA at the discretion of the federal Minister of Environment in order to avoid two EAs. Public participation has been retained, and Section 24 states the government authority responsible must ensure the public has an opportunity to participate in the EA process.

However, the revisions embodied in CEAA 2012 restrict the "environmental effects" to be considered and limit the number of projects subject to an EA. As a result, there will be fewer EAs with a narrower scope, shorter timelines for decisions, fewer federal agencies and departments involved, more discretion in the decision-making process, a smaller number of persons given "interested party" status, and increased federal willingness to substitute provincial EAs than was previously the case. The extent to which the federal focus on increasing the speed and efficiency of the EA process will undermine the strategic planning value of federal EAs in mitigating environmental, social, and cultural impacts remains to be seen.

The constitutional division of jurisdictional authority between the federal government and the provinces has engendered legal interpretations and political debates that have in many ways defined Canada's history. This political history is not just about who has authority – it has also been about the question of who benefits and who pays. Technically, the federal government has constitutional jurisdiction for energy development in Canada's northern territories and coastal zone. But two provinces also have lands in the northern coastal zone (BC and Newfoundland and Labrador), and federal-provincial arrangements will be required to address many of the logistical and economic aspects of northern energy development. But the question of who benefits and who pays will continue to drive much of the discussion. The possible answers become more complex when other significant actors in northern energy development are considered: specifically, billion dollar energy corporations and Aboriginal peoples.

Between 2003 and 2012, 25 public oil and gas firms operating in Canada achieved the billion dollar annual revenue status, and 22 are based in the province of Alberta (Centre for Digital Entrepreneurship and Economic Performance 2014). The long-term strategic influence of companies in this revenue bracket and the economic and political leverage that such companies are capable of exerting internationally as well as on the federal government and the provinces makes them critical to future northern energy development.

Aboriginal peoples (defined as First Nations, Metis, and Inuit) also have rights protected in Canada's constitution. Historical treaties made with the British Crown cover large areas of land within Canada's provinces and territories. Aboriginal people and lands fall under federal jurisdiction and a variety of federal legislation including the Indian Act 1876 and 1951 and more recently, legislation finalizing self-government and comprehensive land claim settlements such as the 1984 Cree-Naskapi (of Quebec) Act, the 1993 Nunavut Land Claims Agreement, and the Nisga'a Final Agreement Act in 2000. The federal government has a legal duty to consult with and accommodate Aboriginal stakeholders around any activities affecting their interests. This duty to consult requirement has been contentious in specific situations and has the potential to significantly delay federal government energy, mining, forestry, and transportation project approvals. This issue has recently been played out in Northern British Columbia on Canada's west coast related to LNG tidewater port development proposals. Similarly, a proposed Mackenzie Gas Project in Canada's Northwest Territories would develop an estimated 55 trillion cubic feet of natural gas reserves in the Mackenzie Delta and Beaufort Sea and involve aboriginal interests (Taylor et al. 2010). The legal status of Aboriginal rights was tested before the Canadian Supreme Court in Tsilhqot'in Nation vs British Columbia (2014), and the result affirmed an underlying Aboriginal interest in traditional lands and the Crown's duty to consult.

The status and social and economic condition of Aboriginal peoples in Canada have seen some degree of controversy for most of the country's 150-year history. The Aboriginal Economic Benchmarking Report (National Aboriginal Economic Development Board 2012) identified that there is a significant difference in high school completion rates between Aboriginal (56%) and non-Aboriginal Canadians (77%). The Aboriginal high school completion rate drops even further (40%) for those living on Federal Indian reserves. This difference is especially problematic given the established link between education level and employment success. For example, an analysis of workforce opportunities in the clean energy sector in British Columbia (Globe Advisors 2012, p. 56) found inflated educational requirements for jobs, and a lack of educational upgrading and technical training opportunities in remote areas was a major barrier to energy transition. Similarly, Eco-Canada's 2010 report "Profile of Canadian Environmental Employment" found 39% of professionals in clean energy-related employment required at least a bachelor's degree level of education.

This link between education level and employment puts northern energy development at risk of reinforcing a long-standing historical grievance that "Aboriginal Canadians have not benefited from natural resource development in their traditional territories to the same degree as non-Aboriginal Canadians" (Eyford 2013, p. 22). However, energy development also presents an opportunity to create wealth and significantly improve educational success and quality of life for the young and growing Aboriginal demographic in the northern coastal zone, but only if they are able to participate in a significant and culturally appropriate way.

Sustainable Energy Mix Driving Forces

Four key issues/driving forces are already at work and will play a primary role in determining the future of Canada's "far north" coastal zone:

- The geopolitics of energy and mining market demand and pricing;
- Rapidly changing climatic conditions affecting artic sea ice and seasonal terrestrial temperatures "north of 60";
- The predominance of First Nations, Metis, and Inuit populations and territories in Canada's north and northern coastal zone;
- Access to financing for new and large-scale energy and transportation infrastructure.

In addition to these larger forces, remote community energy transition away from diesel dependency needs to be supported by a critical subset of more immediate and local needs that are not unique to cold climates but have been found to apply to energy transitions in a variety of remote community and island locations (Niezl 2010):

- Access to broadband
- Access to technical expertise, education upgrading, and employment training
- Cost of hydrocarbon-based fuel and electricity
- Community support
- Access to appropriate technology
- Committed local and external financial resources

The World Energy Council (2012, p. 3) defines "energy sustainability" as consisting of three core dimensions – "energy security, social equity, and environmental impact mitigation" which are dependent upon "… complex interwoven links between public and private actors, governments and regulators, economic and social factors, national resources, environmental concerns and individual behaviours" (The World Energy Council 2012, p. 3). In the daily life of Canada's northern coastal communities, community and family social organization, individual and collective life ways, and sociocultural meaning and experiences are all tied to living and surviving collectively in a remote and fragile environment under extreme climatic conditions. Both Aboriginal and non-Aboriginal populations in remote northern communities understand the value and scale of the potential energy and mining developments coming their way (Eyford 2013). However, they also understand that their future and their children's future will depend on achieving long-term energy sustainability. For people in Canada's northern coastal zone, there are no trade-offs between energy security, social equity, and environmental impact mitigation.

The term "chilly climate" which literally describes Canada's Arctic territories also seems to describe the historical federal government policy environment for renewable energy development in northern communities. For example, the off-grid working group of the Renewable and Electrical Energy Division, Energy Policy Sector, Natural Resources Canada (2011, p. 38) concluded "…little has been done to integrate local resources into the energy mix of these communities." Similarly, a business case analysis for renewable energy in four remote northern communities clearly identified the primary barriers to renewable energy and energy sustainability as "limited financial capacity, limited ownership, limited human capacity, logistics in remote communities" (CTCG 2012, p. 2). Previous federal governments have stated clear preferences for devolving power to the northern territorial governments and relying on competitive markets to take care of energy needs. However, there still is a critical jurisdictional and political role for the federal government to play in fostering collaborative governance and developing renewable energy transition strategies and a sustainable energy mix in Canada's north.

Lessons Learned, Information Gaps, and Priorities for Action

A number of lessons about renewable energy and energy mix can be learned from a variety of international experiences in remote and fragile environmental contexts. A review of this literature identifies some common factors emerging from a variety of different circumstances. These findings clearly identify the importance of both community and national government roles. Specifically, community involvement strategies engaging local stakeholders in all project phases from planning to operation and maintenance can maximize local benefits including the development of local expertise, employment creation, and local business development.

Successful projects have involved supportive institutional relationships between remote area governments and national governments related to "... utility structure and regulation, electricity market structure and conditions" (IEA-RETD 2012, p. 91). In addition, community projects have been successful when national governments have:

- Adapted policies or created policies specific to renewable energy development in remote areas;
- Enabled new institutional and regulatory frameworks;
- Created alternative and location-specific tariff structures (such as FIT);
- Enabled location-specific integrated utility systems for hybrid power systems.

The United Nations Framework Convention on Climate Change's affiliated 2004 International Conference on Renewable Energy Policy Conference in Bonn, Germany (Canadian Renewable Energy Alliance 2006), identified one major lesson learned as the importance of integrating renewable and nonrenewable energy sector policies with specific targets, strategies, timelines, and implementation plans for energy mix end users. A second lesson identified was the importance of increasing and building public awareness, knowledge, and support for the transition to renewable energy (Canadian Renewable Energy Alliance 2006). The importance of revising local building codes to accelerate the use of renewable technologies and the importance of licensing and siting in enabling renewable energy deployment were also identified as international lessons learned.

The results of a comparative study on rural and remote electrification policies in emerging economies (OECDS/IEA 2010, p. 99) identified a number of "preconditions" for successful renewable energy deployment in rural and remote environments:

- Good data gathering and community planning to understand spatial distribution of energy requirements and uses to identify energy targets for selecting suitable technologies;
- Sustained support from government and long-term funding commitments. Independent and dedicated institutional and management structures;
- Secure market infrastructure to attract private investment;
- Community involvement throughout the decision-making process.

The significance of "community involvement" relates to including local knowledge related to energy consumption patterns and local conditions affecting technology transfer. It also includes the advantage community members have in understanding community protocols and social relationships. Community members are more likely to respond to and work with other respected community members, extended family, and intergenerational connections than they are to outside experts. Community engagement is key to whether or not "social license" exists. Finally, a key lesson learned is the importance of risk and risk perception in financing renewable projects remote and fragile environments.

Perhaps the most important information gaps affecting energy mix potential in remote and fragile environments are specific knowledge about Canada's 292 remote northern communities. Renewable energy projects have proven to be most effective when customized to fit local community energy use characteristics and siting situations. However, very little specific information currently exists, and this information is needed to select the appropriate type of technology that best fits community circumstances. A second significant information gap is the strategic and operational implications of northern climate and environmental change. As discussed earlier, seasonal temperature variability, extreme weather events, seasonal changes in sea ice, and terrestrial permafrost conditions all have the potential to affect renewable energy technologies and related community and regional infrastructure. How these changes will manifest and where these changes are most likely to occur and at what scale in the future is not well understood.

In an area as vast as the Canadian northern coastal zone, scale is a primary consideration given the distances involved. Given this large geographic area, the best approach to achieving energy sustainability may or may not be by developing renewable projects one community at a time. Specifically, at the scale of the area involved, there may be other regional or subregional options to "scale up" infrastructure or create scale-free networks to connect more than one community or connect industrial and community energy needs. The possible strengths, weaknesses, opportunities, and threats (SWOT) of scale needs to be much better understood in energy planning and development. Strategies selected to address the scale in energy planning will have a direct effect on infrastructure costs and timelines.

The potential for future oil and gas development in Canada's northern coastal zone and northern territories brings with it the potential to create renewable energy opportunities. For example, "Between 2000 and 2010, U.S. based O & G companies invested roughly $9 Billion in renewables (wind, solar, biofuels) – roughly 20% of the total U.S. renewable investment of $47 Billion over the same period" (T2 and Associates 2011 in Switzer et al. 2013, p. 1). More information is required to better understand the potential drivers for engaging the oil and gas companies with interests in the northern coastal zone to play an active role in northern energy transition and energy mix. Renewables are moving into industrial applications such as solar power for desalination and in the mining industry where operations are often energy intensive and in remote locations. The potential for linking domestic and industrial energy mix may have advantages at specific scales and in specific locations.

In addition to the information gaps identified above, there are also institutional gaps which affect renewable energy development. A SWOT analysis of Canadian energy and climate policies by Fertel et al. identified an absence of a consistent federal strategy, a lack of coordination between energy and climate policy, and a lack of a national renewable energy strategy as current institutional gaps and concluded "There is no coordinated national energy strategy beyond reliance on market forces…" (Fertel et al. 2013, p. 1148).

The dominance of diesel-fueled electricity (86%) in remote northern communities makes establishing the business case potential of specific renewable technologies for displacing diesel and other forms of hydrocarbon-based electricity a priority for action. Demonstration research in four Canadian remote communities suggests a business case can be made based on different renewable sources (CTCG 2012). A second priority for action is to identify the remote communities or areas of interest for renewable projects and begin doing the preliminary data collection, consultation, and feasibility studies necessary to underpin future project development. The types of studies and information necessary to proceed with project development can take several months if not years and can delay project construction if not started well in advance of timelines for project construction and operation. A third priority for action is to initiate community-based training and educational upgrading to maximize employment opportunities in renewable energy project construction, operation, and maintenance. Lead time is necessary to ensure that once a system is installed, the human resources are in place to ensure its performance.

Given the impact of risk perception on financing, the fourth priority for action is to better understand this problem and work with investors and other stakeholders to identify possible financial tools and other risk management mechanisms to facilitate remote renewable energy project financing.

Conclusion

A number of enabling factors are necessary to ensure renewable energy projects are successful. While choice of an appropriate technology is certainly one of them, it is not the only one. Technology is a necessary but not sufficient consideration for achieving a sustainable energy mix in remote and fragile environments.

Canada's northern remote communities need and will increasingly need access to a wide range of engineering and technical expertise in order to achieve energy sustainability. Indeed, if the predictions of the "Renewables 2013 Global Status Report" (REN21 2013) are any indication, the demand for technical expertise is going to grow significantly. For example, the renewable energy contribution to global energy is generally estimated to be in the 15–20% range. However, the 2013 Global Status Report predicts (based on actual 2011 GW capacity data) that renewables will be making a far different contribution to global energy (REN21 2013, p. 20):

- Wind power capacity increasing between 4-fold and 12-fold [4-fold or 12-fold means 4 times more or 12 times more];
- Solar PV [photovoltaic] between 7-fold and 25-fold;
- CSP between 20-fold and 350-fold [Concentrating Solar Power plants use mirrors to focus sunlight into a light beam used to heat fluid to drive a steam turbine to produce energy];
- Bio-power between 3-fold and 5-fold;
- Geothermal between 4-fold and 15-fold; and;
- Hydro between 30% and 80%.

Clearly, achieving such transformational numbers will require significant technological expertise and innovation. But, it also will require innovation and expertise in financing, markets, institutional frameworks, regulatory regimes, planning, management, public education, and a number of other specialized bodies of knowledge and skill sets.

The Canadian context as presented here does not focus on wind-fuel cell capacity or hydrogen electrolyzer performance as the critical factors affecting the renewable energy future of remote northern coastal zone communities – although this level of technical detail is certainly important in a specific project context. Sustainable energy mix as presented here as an integrative process that operationally integrates renewable energy technologies within scale-specific social, economic, and biogeoclimatic contexts. This contextual and integrated approach to energy mix design is important because the definition of energy sustainability is context dependent.

International cooperation can also make an important contribution to achieving sustainable energy mix and energy sustainability. It is critical to share knowledge and experience from a variety of perspectives and experiences about what works and what does not. It is also important to identify international research priorities and problem-solving opportunities, including joint venture demonstration projects to understand the interdisciplinary, cross-cultural, environmental, and technical challenges involved in a variety of circumstances. Understanding how to design, plan, and manage for a sustainable energy mix is critical to the future of Canada's northern coastal zone. But, it is also critical to the future of the world's remote and fragile environments.

References

Arctic Council (2013) Arctic Resilience Interim Report 2013. Stockholm Environment Institute and Stockholm Resilience Centre, Stockholm. www.arctic-council.org/arr

Arctic Marine Shipping Assessment (2009) Report. Arctic Council, April 2009, second printing. https://www.pmel.noaa.gov/arctic-zone/detect/documents/AMSA_2009_Report_2nd_print.pdf. Accessed June 2014

Benidickson J (2013) Environmental law, 4th edn. Irwin Law Inc., Toronto, ON. www.irwinlaw.com

Bird KJ, Charpentier RR, Gautier DI, Houseknecht DW, Klett TR, Pitman JK, Moore TE, Schenk CJ, Tennyson ME, Wandrey CJ (2008) Circum-arctic resource appraisal: estimates of undiscovered oil and gas north of the arctic circle. USGS Fact Sheet 2008–3049. U.S. Department of Interior, U.S. Geological Survey. Menlo Park, CA. https://pubs.usgs.gov/fs/2008/3049/fs2008-3049.pdf. Accessed June 2014

Canadian Renewable Energy Alliance (2006) Framework for a model national renewable strategy for Canada. Accessed at https://www.pembina.org/pub/1278

Centre for Digital Entrepreneurship and Economic Performance (2014) Canada's billion dollar firms: contributions, challenges and opportunities. http://deepcentre.com/billiondollarfirms

Clean Tech Community Gateway (CTCG) (2012) Planning for Prosperity, positioning remote communities to benefit from clean energy. Westbrook Mall, Vancouver, BC (www.CTCG.org) http://ctcg.org/wp-content/uploads/2013/03/Planning-for-Prosperity.pdf

Dawson C (2014) Oil giants set their sights on arctic waters. Wall Street J Rep, May 18. http://online.wsj.com/news/articles

de Jager D, Rathmann M (2008) Policy instrument design to reduce financing costs in renewable energy technology projects. Prepared for the International Energy Agency, Renewable Energy Technology Development, Ecofys International BV, Utrecht

Eco-Canada (2010) Profile of Canadian environmental employment. Labour Market Research. Environmental Careers Organization of Canada.(www.eco.ca) http://www.eco.ca/pdf/Profile-Of-Canadian-Environmental-Employment-ECO-Canada-2010.pdf

Environment Canada (2012) Progress report of the federal sustainable development strategy, Annex A: clean air agenda. Government of Canada, Ottawa. http://www.ec.gc.ca/dd-sd/23E4714EB774-4CC5-9337-F87B01556727/2012_Progress_Report_ofthe_FSDS.pdf

Eyford DR (2013) Forging partnerships building relationships, aboriginal Canadians and energy development. Report to the Prime Minister. Catalogue No. M4–109/2013E–PDF. Ministry of Natural Resources, Government of Canada, Ottawa

Fertel C, Bahn O, Vaillancourt K, Waub J-P (2013) Canadian energy and climate policies: a SWOT analysis in search of federal/provincial coherence. Energy Policy:63, 1139–1150. Elsevier. www.elsevier.com/locate/enpol

Globe Advisors (2012) Powering our province: an analysis of the clean energy business and workforce opportunities for communities in British Columbia. www.globeadvisors.ca

Government of Canada (1867) Constitution Act, 1867 section 109. http://laws-lois.justice.gc.ca/eng/Const/page-6.html

Government of Canada (1982) Constitution Act, (1982). http:\\laws-lois.justice.gc.ca/eng/acts/Const/page-6.html

Government of Canada (2012b) Jobs, growth and long-term prosperity act. http://laws-lois.justice.gc.ca/eng/acts/J-0.8/

Government of Canada (2014) Northwest territories devolution act. http://www.aadnc-aandc.gc.ca/eng/1100100036087/110010036091

Hausner VH, Fauchald P, Tveraa T, Pedersen E, Jernsletten J-L et al (2011) The ghost of development past: the impact of economic security policies on Saami pastoral ecosystems. Ecol Soc 16(3):4. https://doi.org/10.5751/ES-04193-160304

IEA-RETD (International Energy Agency—Renewable Energy Technology Development) (2012) Renewable energies for remote areas and Islands (Remote). Final Report. http://iea-retd.org/archives/pubications/remote

Marine Renewables Canada (2013) Annual report. http://www.marinerenewables.ca/wp-content/uploads/2012/11/Marine-Renewables-Canada-Annual-Report-2013.pdf

Muir MAK (2013) Using renewable energy and desalination for climate mitigation and adaptation in small island developing states and coasts of arid regions. Outreach On-line Magazine. Stakeholder Forum for a Sustainable Future. http://www.stakeholderforumlorg/sf/outreach. Accessed 2013

National Aboriginal Economic Development Board (NAED) (2012) Aboriginal economic benchmarking report. http://www.naedb-cndea.com/reports/the-aboriginal-economic-benchmarking-report.pdf

National Energy Board (2013) Canada's energy future 2013: energy supply and demand projections to 2035" Cat. No. NE2–12/2013E–PDF. Government of Canada, Ottawa. https://www.neb-one.gc.ca/nrg/ntgrtd/ftr/2013/2013nrgftr-eng.pdf

Natural Resources Canada (2011) Renewable and electrical energy division, Policy Sector. Status of Remote/Off-Grid Communities in Canada. http://publications.gc.ca/site/archiveearchived.html?url=http://publications.gc.ca/collections/collection2013/rncan-rrcan/M154-71-2013-eng.pdf

Natural Resources Canada (2014) Energy policy. http://www.nrcan.gc.ca/energy/energy-resources/15903

Niezl A (2010) Comparative study on rural electrification policies in emerging economies: Keys to successful policies. Information Paper. EA Energy Technology Policy Division, Organization for Economic Cooperation and Development (OECD)/International Energy Agency (IEA), Paris. http://www.iea.org/publications/freepublications/publication/rural_elect.pdf

Nong H (2011) Arctic energy: pathway to conflict or cooperation in the high north? J Energy Security. http://www.ensec.org/index

ONSC 609 (2011) CITATION: Hanna v. Ontario (Attorney General), 2011 ONSC 609 DIVISIONAL COURT FILE NO.: 491/09. DATE: 20110303 www.aph.gov.au/DocumentStore.ashx?id=17f924a3-5a9c-452e-a2ba

Ontario (2007) Endangered species act. https://www.ontario.ca/laws/statute/07e06

Ontario (2009) Green energy act. https://www.ontario.ca/laws/statute/09g12

REN21 (2013) Renewables 2013 global status report, renewable energy policy network for the 21st century. REN21 Secretariat. Accessed at http://www.ren21.net/Portals/0/documents/Resources/GSR/2013/GSR2013_lowres.pdf

Solar Energy (2014) Solar power plants information centre. http://solarenergypowerplants.blogspot.ca/2010/09/sarnia-photovoltaic-pv-power-plant.hi. Accessed June 2014

Statistics Canada (2014) Government of Canada. http://www.statcan.gc.ca/start-debut-eng.html. Accessed Mar 2014

Supreme Court of Canada. Tsilhqot'in Nation v. British Columbia (2014) SCC 44. Date: 20140626. Docket: 34986. https://scc-csc.lexum.com/scc-csc/scc-csc/en/item/14246/index.do

Switzer J, Lovekin D, Finigan K (2013) Renewable energy opportunities in the oil and gas sector, Executive Summary. Pembina Institute. http://www.pembina.org/pub/2411

T2 and Associates (2011) Key investments in greenhouse gas mitigation technologies from 2000 through 2010 by energy firms, other industry, and the federal government. http://www.api.org/~/media/Files/EHS/climate-change/2013-key-investments-ghg-mitigation.pdf

Taylor A, Grant J, Holroyd P, Kennedy M, Mackenzie K (2010) At a crossroads, achieving a win-win from oil and gas developments in the north west territories. Pembina Institute. https://www.pembin a.org/reports/crossroads-nwt-oil-and-gas-revenue.pdf

United Nations, Department of Economic and Social Affairs, Population Division (2017) World population prospects: the 2017 revision, key findings and advance tables. Working Paper No. ESA/P/WP/248

World Energy Council (2012) World Energy Trilemma: 2012 sustainability index. https://www.worldenergy.org/wp-content/uploads/2013/01/PUB_2012_Energy_-Sustainability_-Index_VOLII1.pdf

World Population Review (2017) http://worldpopulationreview.com/states/california-population/. Accessed Mar 2017

Chapter 11
Sustainable Energy Mix in Fragile Environments: A Transdisciplinary Framework for Action

Mary-Ellen Tyler and Irene M. Herremans

Introduction

The concept of "energy mix" is a function of context, scale, and energy source availability. To date, renewable energy sources (such as wind, solar, micro hydro, and biofuels) have been viewed as technical solutions primarily for local and regional needs although renewable energy production has the potential to be expanded at much larger scales and in much larger markets. Regardless of the type of energy source, there are always a variety of contextual, temporal, scale, social, cultural, economic, environmental, and technological factors affecting source selection, production, distribution, and use and price. A long-term sustainable approach to designing energy mix in fragile environments needs to consider the strategic and contextual factors affecting energy source choices, limitations, and opportunities. Therefore, understanding energy mix alternatives in the context of dynamic interactions among social, economic, and environmental systems is key to implementing sustainable energy mixes in fragile environments. A transdisciplinary approach involves integrating knowledge from multiple disciplines and the creation of new knowledge through social learning and partnerships among researchers, professionals in government and private sector, community leaders, and local residents.

M.-E. Tyler (✉)
Faculty of Environmental Design, University of Calgary, Calgary, AB, Canada
e-mail: tyler@ucalgary.ca

I.M. Herremans
Haskayne School of Business, University of Calgary, Calgary, AB, Canada
e-mail: irene.herremans@haskayne.ucalgary.ca

Energy Mix as a Wicked Problem

Ultimately, energy and energy mix choices involve decision-making informed by available knowledge and guiding values (Balint et al. 2011). The extent of available knowledge and agreement on guiding values is highly variable depending upon context. This contextual variability results in what Balint et al. (2011) have referred to as a decision-making continuum ranging from structured to poorly structured and unstructured problems. The degree of available empirical knowledge and shared values is usually highest in well-structured problem situations and minimal to absent at the other end of the continuum. The significance of this continuum in both public and private sector decision-making is that there is increasing uncertainty between actions and outcomes in unstructured situations relative to highly structured. Specifically, energy mix may appear to be a well-structured problem. But, in practice, it is often characterized by little to no agreement about guiding values, limited empirical knowledge or uncertain science, and no clear relationship between alternative actions and outcomes. As such, energy mix is a "wicked problem," and wicked problems are characteristic of complex systems behavior characterized by surprise, uncertainty, and change over time (Rittel and Webber 1973). Social ecological systems are complex systems. Energy mix is frequently viewed as strictly a function of technology and basic economic principles, insofar as the goal is a cost-effective mix of available energy technologies needed to meet demand. However, it is much more complicated than this even at a small scale. Therefore, recognizing energy mix as a wicked problem can assist in identifying important context variables and interrelationships across a range of scales.

Avoiding the "Logic of Failure"

The inherent uncertainty of policy and management decision-making inherent in complexity is further compounded by a phenomenon documented by Dorner (1996) and referred to as "the logic of failure." This phrase refers to well-intended rational planning and decision-making processes by qualified experts that resulted in counterintuitive and unanticipated outcomes that actually made the original problem worse. Using a number of case studies, Dorner (1996) attributes the unanticipated results to inadequacies in understanding the complexity of the systems involved. As a result, Dorner (1996, p. 199) recommends "systemic thinking" to frame goals, decisions, and actions in a dynamic systems context to better understand the critical interrelationships involved. In order to better deal with system complexity, Dorner (1996, p. 79) identifies three types of knowledge requirements: "We need to know on what other variables the goal variables that we want to influence depend. …;" "We need to know how the individual components of a system fit into a hierarchy of broad and narrow concepts. …;" and "We need to know the component parts into which the elements of a system can be broken and the larger complexes in which those elements are embedded."

As with wicked problems, it is uncertainty and "surprise" that contribute to the "logic of failure." Experienced decision-makers and experts working from knowledge gained from certainty in well-structured situations are too often ill prepared to deal with unstructured situations lacking empirical knowledge and with great uncertainty about relationships between actions and outcomes. To address the problem of energy mix in fragile environments, it is important to frame energy mix as involving complex systems behavior and not assume there is a high degree of certainty between action and outcome.

Fragile Environments as Social Ecological Systems

Fragile environments are of interest primarily because of their geographic locations. Limited accessibility has enabled biologically and geologically significant features and indigenous or historically unique sociocultural populations to persist. Such venues often become desirable for tourism or resource extraction activities but lack the technological infrastructure and financial resources found in more developed and urbanized contexts. Generally, fragile environments occur in more rural and remote locations including islands, mountainous areas, and extreme climatic conditions such as deserts, rainforests, and frozen tundra. In some instances, they are large protected areas such as national parks, nature reserves, and World Heritage sites. Regardless of the specifics, every fragile environment exists in a larger context. It is within this larger biogeoclimatic and geopolitical context that driving forces and variables including technology, transportation, economic value, legal, and institutional frameworks impinge on energy demand and supply.

Energy Mix as a Sustainable Development Strategy

One of the primary challenges to energy mix in fragile environments is human population growth and related development activities which are accompanied by increasing energy demands. Increasing population and development pressures associated with resource extraction and ecotourism puts pressure on existing social and ecological systems to provide more food, water, and energy. While a steady-state economy may be appropriate for some developed economies, many developing economies have yet to reach a stage where they can concentrate more on developing quality of life. The social and environmental costs of growth in some cases can make growth unsustainable, although this has seldom stopped growth or changed political decisions. Energy mix can be an important part of a sustainable development strategy but is seldom considered.

Rural and remote locations generally have relatively small populations, limited technical capacity, and limited access to financial capital. As such, remote location, limited access, and limited capital, have historically mitigated against the development of conventional energy supply and infrastructure, which in turn has constrained development potential. In order to take advantage of social and economic development opportunities, it is necessary to consider linking existing social, economic, technical, and ecological resources to function as an interdependent and mutually reinforcing social ecological system. There is no room for trade-offs between social, economic, and ecological resources in fragile environments when resources are limited. Rather, the "art" of energy mix design involves recognizing and understanding the key social ecological system linkages necessary to ensure sustainable and effective energy technology transfer.

Galapagos Energy Mix Themes

Historically, the primary source of energy in the Galapagos has been imported petroleum. Despite Ecuadorian government subsidies, petroleum supplies to the Galapagos remain relatively expensive. For example in 2013, the subsidized price of gasoline was $1.50 (USD) a gallon and diesel-based electricity without subsidy more than 40 cents/kwh (Finley 2013). The use of imported petroleum also has environmental costs in the form of air quality and risks of marine spills not factored into the cost. A transition away from fossil fuels has been underway in the Galapagos to address both cost and marine environmental concerns. For example, since 2007, renewable energy use on San Cristóbal Island has replaced 30% of petroleum-generated electricity. This is equivalent to approximately 8.7million liters of diesel fuel and 21,000 tons of carbon dioxide emissions, according to project sponsors Global Sustainable Electricity Partnership (Vella 2016).

In the context of this transition, five themes affecting energy mix emerged at the World Summit on Sustainable Energy in Fragile Environment held in San Cristóbal in July 2014:

- Need for new institutional frameworks
- Fossil fuel subsidy dependency
- Importance of social capacity
- Potential capacity of renewables
- Receptivity to technology transfer

These themes are consistent with the typology of mix factors identified in Fig. 11.1 and described in more detail as follows.

1. *Need for New Institutional Frameworks*
   Current regulatory systems in most countries favor fossil fuels, which in turn, create entry barriers for other energy options. Subsidies, when implemented responsibly, can help shift market receptivity and encourage new energy mix options. However, markets need to reflect the true cost of energy, which means eliminating subsidies to unsustainable energy sources. Incentives to move toward energy sources and technologies with less social and environmental costs can help motivate market change. It is also important to consider strategies and timelines for changing subsidies to reduce financial impacts on specific sectors or income groups. While governments play a

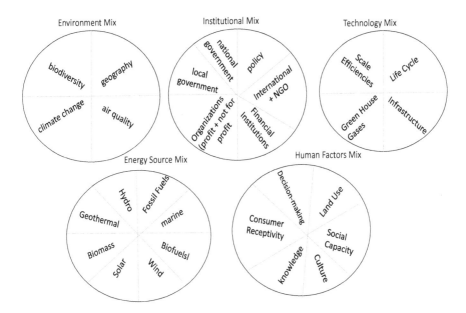

Fig. 11.1 Five energy mix typologies

major role in providing public policy direction, political stability as policy implementation problems create obstacles for energy mix solutions (Hoffman et al. 2008). Similarly, energy policy driven by political lobbyists, special interests, or subsidies in exchange for votes, can also deter the development of sustainable energy policy.

Breaking down institutional and organizational "silos" is critically important to enable greater integration of social, economic, and environmental goals in energy mix planning. New institutional frameworks are necessary for engaging with complex problem-solving and for dealing successfully with sustainable energy policy development. Similarly, determining an appropriate energy mix involves a cross-sectoral and collaborative process. Institutional frameworks necessary to support sustainable energy mix development need to include the following components:

- Policy connections at multiple levels
- Participation and engagement of local governments and communities
- National or international availability of resources and expertise not available locally
- Financial institutions or corporate partners with financial resources
- Provision of skilled labor to install, operate, and maintain facilities and related infrastructure

Both public and private financial institutions can play important roles in providing funding and investment innovations to enable businesses to move to commercialization of alternative energy options with some arrangement for shared risk taking. Innovations can have difficulty moving from the research and development stage to commercialization because of risk. Even if corporations are ready to purchase new technologically feasible innovations and banks are ready to fund commercially feasible innovations, the challenge for most entrepreneurs is to demonstrate both technical and commercial feasibility in order to get needed funding. Governments can assist in sharing the risk involved in moving innovations through these critical feasibility phases. Organizations at the community level and local governments also need to take responsibility for finding sustainable solutions for their own energy needs. Communities need to be encouraged to share knowledge, experiences, and ideas as part of a "top-down" and "bottom-up" approach to move the energy mix transition toward more sustainable strategies.

2. *Importance of Social Capacity*

 The idea of social capital comes from experience with community development. Social capital reflects stable functional and value-based relationships. Households, neighborhoods, and communities characterized by trust and cooperation have strong social networks based on working together and sharing resources (Nahapiet and Ghoshal 1998). Social learning is an important part of social capital and involves the sharing of ideas and collaborative learning. Similarly, social capacity contributes to solving energy problems at many levels. Partnerships with nonprofit organizations, local governments, communities, financial institutions, research centers, universities, and private and public sector organizations can increase knowledge and support for alternative and sustainable energy choices, as well as identify the resources and expertise required for implementation of new projects. Social capacity building includes all age groups. Access to new information, education, and skills training can create new employment opportunities and an informed intergenerational public that is cognizant of energy mix issues and choices.

3. *Fossil Fuel Subsidy Dependency*

 Fossil fuel dependency has resulted in significant institutional regulation. Fossil fuels drive the economies of most countries and represent a large percentage of gross national product (GNP). As the dominant international energy source, fossil fuels are profitable for both the private and public sector. Public sector revenues from fossil fuels subsidize social services and infrastructure and create a higher standard of living. The production, transportation, and consumption of fossil fuels and their by-products employ large numbers of people. Understandably, historical investment in and socioeconomic benefits accruing from fossil fuel use creates resistance to change. As is the case with many public policy issues, the status quo is often preferred because it offers certainty. Alternatively, a change in energy sources is a risk and perceived as having high transaction costs, uncertain consumer receptivity, and uncertain technical and organizational implications (Stern 2006; Wong et al. 2014). There is no "perfect" energy source free of consequences. Every energy source has pros and cons which makes justifying the subsidy status quo of fossil fuels easier to maintain. Fossil fuel subsidies create economic advantages for fossil fuels that give them a competitive edge that is very hard to overcome. If fossil fuel subsidies remain unchanged, then similar subsidies for alternative energy sources become necessary in order to create some kind of level playing field. Rather than artificially creating preferred energy choices though public subsidy, a sustainable energy mix must reflect the merits of each available energy source in its operational context. Economic subsidies can be tools to achieve social and ecological benefits, but they also create economic advantages for preferred interests independent of social and ecological consequences. The politics of energy is clearly a reality, but a level policy field that enables energy mix development based on operational and contextual merits is critical for fragile environments.

4. *Potential Capacity of Renewables*
   Fossil fuel dependency is in large part due to its operational history in providing predictable results across a range of scales, contexts, and uses. Alternative energy sources have yet to demonstrate the same operational capacity in many countries. The future of renewable energy sources is dependent on their ability to compete with fossil fuels so there needs to be certainty about performance efficiencies across similar scales and situations. The transition to alternative energy sources and changes in energy mix is dependent on successful evidence-based demonstration. For example, the Renewables 2016 Global Status Report (REN21 2016, p. 17) stated that by 2015 renewable energy sources had become accepted mainstream energy sources worldwide. A number of factors accounted for this operational certainty including "cost-competiveness of renewable technologies, dedicated policy initiatives, better access to financing, energy security and environmental concerns …" and "new markets for both centralized and distributed renewable energy."

   In the context of the Galapagos, a major renewable transition process has been underway since 2007 as part of the Galapagos Islands Zero Fossil Fuel Initiative sponsored by Ecuador's Ministry of Electricity and Renewable Energy (Carvajal 2012). As an example of renewable capacity, Seymour Airport on the Galapagos Island of Baltra received LEED Gold sustainability certification from the US Green Building Council in 2014 (Velasco 2015). The airport handles approximately 400,000 people annually using 100% renewable energy generated by solar panels and 3 wind turbines.

5. *Receptivity to Technology Transfer*
   In the legal sense, technology transfer means the transfer of hardware and intellectual property between countries and corporations through markets and trade agreements. While access to new energy technology and related intellectual property is critical, technology transfer also has important sociocultural dimensions. For example, Costanza (2000) has described four visions of the future ("star trek, mad max, big government, and ecotopia") which reflect differences in technology dependence as well as social values and behavior. Even the best technology will not be successful if not adopted by consumers. In general, consumers want low prices and supply security, but they also want ease of use and minimal maintenance. Acquiring new skills, knowledge, and confidence in the use of new energy technologies and sources requires both economic and social incentives. Introducing energy education early in public school curriculums can be an important first step. Public school curriculum can assist in building an understanding of the basics of energy sources, energy use, and the importance of sustainable energy solutions.

   Energy studies can also be beneficial in addressing social resistance to change and identifying key factors contributing to success or failure. For example, as part of the "e8" San Cristobal wind and solar projects, a long-distance micro-solar learning program was developed in collaboration with Solar Quest, an international not-for-profit educational organization. Schools had solar installations to provide a power source for new computers and internet connections. Students learned computer science skills in order to learn about energy efficiency and renewable energy technologies with the objective of using these new skills and energy in their communities. Specifically, between 2004 and 2006, students at the Colegio Tecnico Ignacio Hernandez (CTIH) in San Cristobal have participated in the Action, Communications, Technology, and Science Program (ACTS) "… each provided 200 hours of community service to monitor and analyze the Island's electric grid in order to research the potential of reducing electricity demand and consumer energy costs through energy efficiency" (e8 2008, p. 46). To assist with this, "Enel," an e8 Italian company, processed and analyzed the student's data to verify its accuracy (e8 2008, p. 47).

Ecotourism also offers important educational opportunities for informing visitors about the importance of achieving a sustainable energy mix in the Galapagos. The economic impact of tourism and the opinions and choices of tourists and tourism operators have a significant influence on the transition to a sustainable energy mix. Tourism creates part of the increasing energy demand in Galapagos. It requires more energy and infrastructure to support increasing demand for both land- and water-based activities including cruises, land and air transportation, hotels, restaurants, and a large local service sector.

However, while education is important in shaping attitudes and value-based choices, the key to technology transfer will be the customizing of technology use and access to its operational sociocultural operating environment. The typology of mixes illustrated in Fig. 11.1 represents the key factors affecting technology transfer potential. Technology transfer is contextual and needs to respond to different target groups and environmental factors including different levels of wealth, education, geographic location, belief systems, and worldviews. For example, technology transfer capacity in well-educated, wealthy, urban populations in an industrialized country is significantly different than it is in indigenous populations in poor, isolated rural areas. Energy mix design integrates energy sources and technologies with the operational capacities available in the social, economic, institutional, and ecological environment.

The Politics of Energy Mix

The overarching context, in which virtually all factors affecting energy mix exist, incorporates the geopolitics of energy and politics at the national, regional, and local levels. The politics of energy presents a different connotation of "power," and political power often determines who benefits and who pays from energy decision-making. Although energy mix is a wicked problem characterized by uncertainty and a lack of empirical knowledge, when political positions are polarized or intractable, energy mix problems cannot be solved by more research (Keohane and Victor 2013). Domestic politics generally dominates energy policy, and a major change in energy policy is unlikely if there is no political will behind it. However, energy markets and energy-related environmental issues have become increasingly transboundary (Keohane and Victor 2013). International cooperation is often difficult due to historical conflicts and cultural differences. As a result, any effective transboundary or international collaboration is usually a function of multilateral institutions or issue-specific non-state actors. Where common interests align, regions or neighboring countries often cooperate through decentralized networks in the interests of mutual gain (Keohane and Victor 2013). In the context of the Galapagos, a specific example of this type of international collaboration is the e8 international energy development partnership renamed the "Global Sustainable Electricity Partnership" in 2011 (Global Sustainable Electricity Partnership 2016).

Energy mix choices and the transition to alternative energy sources are as much an exercise in the politics of energy as it is a technical exercise. Energy mix is a truly interdisciplinary process that integrates science and engineering with social acceptance, economics, and politics. As Rutherford and Coutard (2014, p. 1353) state: "… energy in a variety of guises is bound up technically, economically and politically with our societies, communities and livelihoods in very diverse ways."

Energy Mix Design in Fragile Environments

Moving to a sustainable energy mix involves more than the installation of wind, solar, and other technologies to produce energy. Energy mix options need to function in specific social ecological operational contexts and need to respond to their specific contexts. For example, Doris et al. (2009, p. 122) identified 14 renewable energy context factors that can have positive, negative, or neutral influences depending upon the specific circumstances. These factors include:

- "Resource availability
- Technology availability
- Technology cost
- Energy costs
- Economic factors
- Project financing options
- Ownership options
- Transmission issues
- Environmental considerations
- Institutional structures
- Land-use issues and constraints
- Information dissemination
- Social acceptance
- Larger policy context"

Energy mix solutions need to incorporate these key contextual factors and address growing energy needs by integrating and strengthening social, ecological, and economic system dynamics. However, how this is done, the nature of the operational and institutional contexts in which the fragile environment of interest exists, and the options available are not always obvious. A systematic and interdisciplinary framework for energy mix design can help identify what factors and energy sources are most important in specific situations. Five different types of "mix" typologies that can be used in identifying, customizing, and designing appropriate energy mix options are illustrated in Fig. 11.1.

Expertise from many different disciplines and sources is necessary in order to evaluate the opportunities and constraints presented by each of these mix types in different fragile environment contexts. For example, sustainable energy mix options in the Galapagos (illustrated in Fig. 11.2) are likely to be different from the mix options for fragile environments found in different biogeoclimatic and geographic locations like Northern Canada. But all sustainable energy mix options for fragile environments will involve some combination of the five mix types in Fig. 11.1.

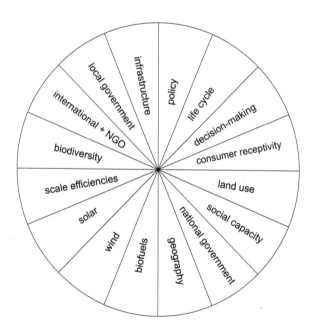

Fig. 11.2 Galapagos sustainable energy mix factors

Conclusion

Moving to a more sustainable energy mix in fragile environments ultimately must address what the World Energy Council (Wyman 2016, p. 6) has referred to as "the energy trilemma":

- *"Energy security*: ... effective management of primary energy supply...reliability of energy Infrastructure, and ability of energy providers to meet current and future demand"
- *"Energy equity*: Accessibility and affordability of energy supply across the population"
- *"Environmental sustainability*: ...achievement of supply and demand-side energy efficiencies and development"

There is no single energy mix capable of meeting all the operational needs of all fragile environments. Energy mix is subject to human values and worldviews, institutional influences, market forces, geographic location, and affordable technology considerations. Changes in energy mix can have far-reaching social, ecological, and economic consequences, so incremental steps are necessary to simulate and demonstrate operational performance in context and at different scales. Developing a sustainable energy mix to meet fragile environment requirements involves partnerships among governments and key public and private stakeholders. As Stirling (2014, p. 83) states: "Understanding possible 'sustainable energy' transformations requires attention to many tricky issues in social theory: around agency and structure and the interplay of power, contingency and practice. These factors are as much shaping of the knowledges and normativities supposedly driving transformation, as they are shaped by them."

The "art" of designing and developing sustainable energy mixes in fragile environments lies in finding an appropriate "mix of mixes" reflecting the specific conditions and operational interrelationships appropriate to specific locational social ecological contexts. A transdisciplinary framework for practice has the potential to develop a mix of practical strategies necessary to address the need for new institutional frameworks, reduced subsidy dependency, increased social capacity, and receptivity to technology transfer in a variety of fragile environment contexts.

References

Balint PJ, Stewart RE, Desai A, Walters LC (2011) Wicked environmental problems. Island Press, Washington DC

Carvajal P (2012) Galapagos Islands Zero Fossil Fuel Initiative. Ministry of Electricity and Renewable Energy-Ecuador. Retrieved from http://www.irena.org/DocumentDownloads/events/MaltaSeptember2012/Pablo_Carvajal.pdf

Costanza R (2000) Visions of alternative (unpredictable) futures and their use in policy analysis. Conserv Ecol 4(1):5

Doris E, McLaren J, Healey V, Hockett S (2009) State of the States 2009: Renewable Energy Development and the Role of Policy. Technical Report NREL/TP-6A2-46667. October 2009. National Renewable Energy Laboratory, U.S. Department of Energy, Office of Energy Efficiency and Renewable Energy

Dorner D (1996) The logic of failure:recognizing and avoiding error in complex situations. Metropolitan Books, New York

e8 (2008) The san cristobal wind and solar projects. Retrieved from http://www.globalelectricity.org/upload/File/e8_san_cristobal_wind_and_solar_projects_publication_final.pdf

Finley R (2013) Energy trends insider. Retrieved on line http://www.energytrendsinsider.com/

Global Sustainable Electricity Partnership (2016) Galapagos San Cristobal Island Wind Project 2003–2016. Retrieved from http://www.globalelectricity.org/upload/File/final_english_galapagos_report_2016(1).pdf

Hoffman VH, Trautmann T, Schneider M (2008) A taxonomy of regulatory uncertainty—application to the European emission trading scheme. Environ Sci Pol 11(8):712–722

Keohane RO, Victor DG (2013) The transnational power of energy. Daedalus J Am Acad Arts Sci 142(1):Winter 2013

Nahapiet J, Ghoshal S (1998) Social capital, intellectual capital, and the organizational advantage. Acad Manage Rev 23(2):242–266. Retrieved from http://w.jstor.org/stable/259373

REN21 (2016) Renewables 2016 global status report. REN21 Secretariat, Paris

Rittel HWJ, Webber MM (1973) Dilemmas in a general theory of planning. Policy Sci 4:155–169

Rutherford J, Coutard O (2014) Urban energy transitions: places, processes, and politics of social-technical change. Urban Stud 51(7):1353–1377

Stern N (2006) Stern review on the economic effects of climate change. Executive summary. HM Treasury, London

Stirling A (2014) Transforming power: social science and the politics of energy choices. Energy Res Soc Sci 1(2014):83–95. Elsevier Ltd

Velasco H (2015) Galapagos airport evolves to renewable energy only. AFP Business Insider (July 18). Retrieved from http://www.businessinsider.com/afp-galapagos-airport-evolves-to-renewable-energy-only-2015-7

Vella H (2016) Galapagos: fragile archipelago switches petrol for renewables. www.power-technology.com, August 31. Kable Intelligence Limited, UK. Retrieved from http://www.power-technology.com/features/featuregalapagos-fragile-archipelago-switches-petrol-for-renewables-4989944/

Wong PSP, Owczarek A, Murison M, Kefalianos Z, Spinozzi J (2014) Driving construction contractors to adopt carbon reduction strategies—an Australian approach. J Environ Plan Manag 57(10):1465–1483

World Energy Council/Oliver Wyman (2016) 6. Retrieved from https://www.worldenergy.org/wp-content/uploads/2016/10/Full-report_Energy-Trilemma-Index-2016.pdf

Index

A
Aboriginal, 36, 83, 165, 167, 173, 174
Aboriginal people, 107, 110, 167, 172, 173
Acidification, 22
Adapting, 26
Adaptive capacity, 157–159
Adaptive radiation, 2
Administrative law, 68, 69
Africa, 58
Agricultural land, 15, 22, 52
Agricultural production, 34
Agriculture, 5, 64, 107, 136, 154, 155
Air compressors, 5
Air conditioning (AC), 40–43, 45, 50
Air emissions, 131, 133
Airplanes, 2, 4, 8, 11, 12, 16, 17
Airports, 4, 22, 30, 167, 189
Air quality, 186
Alternative energies, 15, 16, 29, 50, 77, 96, 102, 153, 187
Alternative energy sources, 26, 103, 188, 189, 191
Amazon region, 107–109, 112, 117, 118, 123, 127, 129, 135–137
Anaerobic degradation, 145
Anaerobic digestion, 153, 154
Annual visitation, 23
Anthropogenic, 141
Antigua, 22
Appliances, 8, 16, 25, 50
Aquifers, 28
Archeological studies, 37
Archipelago, 1, 3, 6, 9, 11, 21, 23, 28, 60, 61, 93, 94

Arctic Circle, 163
Arctic shipping, 164
Artisanal fishing, 5
Assessment method, 94, 137
Asset holder, 38
Atlantic Ocean, 64, 169
Audits, 37, 39–46

B
Bahamas, 22
Baltra, 9, 15, 49, 189
Baltra Island, 11, 30, 50
Barbuda, 22
Barriers, 21–30, 80, 123, 136, 165–167, 173, 175, 186
Baseline, 96–97, 100, 102, 104, 129
Baseload, 58, 66
Bats, 14, 83, 109
Batteries, 15, 109, 111, 113, 119–125, 128, 131, 132, 135
Bay of Fundy, 169
Beaufort Sea, 165, 173
Behavioral change, 45
Belief systems, 190
Belize, 22
Best practices, 41
Biodegradable, 55, 145–147, 150
Biodegradable waste, 145, 146
Biodiesel, 52, 54, 55, 96, 98–99, 102, 103
Biodigester, 153–160
Biodiversity, 21, 23, 25, 29, 107, 158
Biofuel engines, 14, 54
Biofuels, 14, 16, 49–55, 98, 102, 153, 177, 183

Biogas, 34, 69, 82, 84, 142, 144, 149, 153–160
Biogeochemical, 164
Biogeoclimatic, 178, 185, 192
Biomass, 29, 34, 53, 54, 63, 64, 69, 71–73, 76–79, 82, 84, 86, 141, 142, 153–155, 169
Biomaterial, 52
Biosphere reserve, 28
Birth rate, 8
Boats, 2, 4–8, 10–14, 26, 49, 50, 55, 93, 96, 98, 113, 118–123, 131, 137, 138, 168
British Columbia (BC), 65, 66, 165, 168, 169, 173
Brundtland report, 111
Buen Vivir, 24
Building envelope, 42
Bunker fuel, 50
Business case, 175, 177

C
Call for tenders, 62, 76, 80–85, 88
Canada, 28–29, 35, 38, 39, 46, 63–69, 82, 84–86, 97, 98, 163–178, 192
Canada Census, 28
Canadian Arctic, 164
Canadian Environmental Assessment Act (CEAA), 171, 172
Canary Islands, 28
Capacity building, 24, 51, 188
Capital costs, 25, 62, 68, 72, 87
Capital intensive, 97
Capital investment, 61, 86
Carbon dioxide emissions, 14, 169, 186
Carbon emissions, 58
Carbon footprints, 33, 42
Carbon intensity, 34
Carbon neutral, 25
Caribbean, 22, 58, 142, 143
Caribbean Centre for Renewable Energy and Energy Efficiency (CCREEE), 27
Caribou, 164
CARICOM, 22
Case studies, 109, 166, 184
Catalyst, 33–46, 116, 117
Ceiling lighting, 40, 41
Census results, 49
Charging stations, 123
Chile, 76, 77, 79–81, 88
China, 58, 74, 77–79, 154–156, 159
Chloride, 45
Clean energy technologies, 24
Climate change, 3, 22–24, 33–46, 58, 93, 107, 135, 167, 169, 171, 175

Climate variability, 163, 165
Closed-bid auction, 62
CO_2, 28, 34, 41, 98, 100, 102, 103, 122, 159, 170
Coal, 2, 58, 169–171
Coal-powered grid, 63
Coastal wind energy, 69, 87
Coastal zone, 163–178
Cobos, Manuel J., 4
CO_2 emissions, 42, 100, 153
Cold environments, 164
Collaboration, 29, 64, 103, 158, 159, 190, 191
Colonization, 4, 107
Commercial shipping, 164
Common law, 68, 69
Communal services, 109, 136
Community-based, 83, 153–160, 177
Community development, 188
Community involvement, 175, 176
Community planning, 176
Compac Tropha Oil Mill, 51
Complexity, 23, 30, 112, 184
Complex systems, 23, 26, 184, 185
Composting, 141
Concession contract, 34, 35
Conduction, 61
CONELEC, 33, 34, 76, 109, 110, 114
Connectivity, 2–4, 11, 168
Consumer receptivity, 188
Consumers, 12, 26, 27, 29, 46, 58, 62, 73–75, 77, 78, 83, 84, 165, 188–190
Continents, 4, 58, 76, 141, 163, 171
Contract, 34–38, 69, 74, 80–82, 84, 85, 88, 109, 110, 114–117, 134, 166
Core principles, 27
Costa Rica, 60
Cost-benefit, 27, 129, 135
Cost-competiveness, 189
Cost effective, 26, 27, 37, 103
Cost factors, 103
Cost of living, 8, 168
Critical factors, 16, 46, 122, 166, 167, 178
Cross-sectoral, 187
Crude oil, 96
Cruise ship, 49
Cultural diversity, 137
Cultural framework, 4
Cultural values, 163, 167
Culture, 24, 108, 117, 157
Cumulative, 46, 70, 100, 147
Cumulative effects, 49
Current shifts, 23

D

Darwin, 2, 7, 42
Darwin finches, 2, 3
Data collection, 95, 119–120, 124, 177
Decision analysis, 94
Decision-making, 94, 136, 172, 176, 184, 190
Decree 1815, 33
Deer Lake First Nation, 29
Demographics, 93, 173
Demonstration, 65, 96, 99, 103, 177, 178, 189
Demonstration research, 177
Deregulated market, 79
Desalination, 45, 177
Desalinization, 28, 45
Design, 16, 37, 38, 42, 100, 113, 116, 117, 119–129, 178, 186, 190–192
Design parameters, 119
Deutsche Gesellschaft fur Internationale Zusammenarbeit (GIZ), 51
Diesel
    engines, 5, 14, 54, 98
    fuel, 13, 14, 28, 29, 49, 50, 52, 55, 64, 76, 165, 177, 186
    generation, 59, 65, 68
    generators, 21, 28, 54, 65
    oil, 11, 21, 53, 55
Discourse, 6
Disease vectors, 3
Distribution, 2, 27, 30, 34, 36, 37, 43, 66, 70, 108, 111, 113–117, 135, 136, 154, 176, 183
District heating, 64
Divergent adaptations, 1
Domestic energy, 165
Domestic solid waste, 131, 134, 135
Dominica, 22
Drinking water, 107, 108
Driving forces, 21–30, 70, 174–175, 185
Drought, 33, 52
Duty to consult, 173

E

E-8, 24, 190, 191
EcoENERGY, 29
Ecological benefits, 189
Ecological effects, 22
Ecological footprint, 155
Ecological systems, 28, 185
Economic analysis, 121–122
Economic development, 29, 50, 51, 60, 86, 113, 136, 155, 159, 163, 173, 186
Economic factors, 29, 103, 191
Economic impacts, 94, 190
Economic incentives, 62, 170
Ecosystem productivity, 164
Ecotourism, 41, 66, 164, 185, 190
Ecuador, 1, 8, 12–15, 21, 24, 25, 33–46, 49, 54, 55, 66, 68, 76, 86, 93, 96–98, 102, 103, 107, 108, 110, 118, 120, 134, 138, 142, 160
Ecuadorian Amazon, 108, 112, 117, 118, 123, 138
Ecuadorian government, 4, 14, 22, 24, 186
Ecuador Ministry of Agriculture (MAGAP), 51
Ecuador Ministry of Electricity and Renewable Energy (MEER), 15, 50–52, 76, 109, 110
Ecuador National Electricity Board (CONELEC), 33, 34, 76, 109, 110, 114
Ecuador National Environmental Policy, 33
Ecuador National Plan for Good Living, 33
Electrical appliances, 25
Electricity
    consumption, 27, 40, 42, 43
    grid, 37, 59, 69, 82
    storage, 65
Electricity purchase agreements (EPAs), 66
Electric-solar boat, 118, 120, 123
Electric vehicles, 64
Electrolyze, 65, 178
El Hierro, 28
El Nino, 2, 22
El Progreso, 157
Emission factors, 41, 97, 98
Emissions, 14, 28, 29, 33, 41, 42, 58, 63, 94, 96–98, 100–102, 104, 113, 118, 129–133, 137, 141, 151, 155, 169, 171, 186
Employment, 34, 84, 113, 133, 173–175, 177, 188
Employment opportunities, 34, 177, 188
Endangered, 3, 4, 6, 14, 83
Endemic species, 1–3, 16
Energy
    access, 108
    audit, 39–41, 44
    conservation, 29, 41, 42
    demand, 10–12, 28, 39, 42, 45, 46, 55, 97, 98, 102, 136, 155, 167–170, 185, 190
    density, 123
    development, 34–36, 38, 39, 46, 50, 52, 58, 59, 61–64, 67, 69, 71–75, 77, 79, 80, 83, 85–87, 109, 112, 137, 159, 163, 167–175, 177, 191
    efficiency, 27, 39, 44–46, 61, 64, 190, 193
    equity, 193
    forecasting, 170
    intensive, 65, 177

Energy (cont.)
  markets, 54, 58, 112, 191
  mix, 1–17, 21–30, 33, 41, 43, 45, 46, 49–55, 93–104, 163–178, 183–193
  mix alternatives, 96, 183
  mix design, 100, 178, 186, 190–192
  policy, 27, 79, 86, 94, 104, 165, 168, 170, 174, 175, 187, 191
  regulator, 60, 61, 70, 170
  regulatory system, 60
  security, 23, 27, 28, 30, 58, 85, 174, 189, 193
  Smart Fund, 27
  Star program, 43
  systems, 14, 22, 23, 26, 28–30, 94, 95, 112, 119
  transformations, 96, 98, 193
  transition, 25, 26, 30, 166, 170, 173–175, 177
  trilemma, 193
  use, 12, 15, 16, 30, 40, 41, 43, 46, 50, 93, 94, 96, 176, 186, 189
Energy-water nexus, 45, 85
Engineering, 35, 37, 178, 191
Engineering analyses, 37
Engineers, 37, 110, 114
Entrepreneurs, 188
Environmental assessment (EA), 38, 67, 171
Environmental education, 118, 158
Environmental factors, 29, 129, 131, 133, 166, 190
Environmental impacts, 36, 42, 49, 67, 94–96, 109, 129–131, 133, 135, 174
Environmental management plan (EMP), 129, 133–135
Ethnic groups, 108, 155
European Renewable Energy Directive, 52
Evidence-based demonstration, 189
Evolution, 1, 2, 163
Exponential rate, 2
Externalities, 26, 27, 136
Extinction, 2–4

F
Failure, 113, 123–126, 130, 184–185, 189
Fatty acids, 52
Fauna, 1, 4, 21, 24
Federal territories, 164
Feedback, 3, 16, 17, 23, 35
Feed-in tariff (FIT), 34, 62, 68–69, 75, 76, 82, 86, 168
Feedstock, 51, 52, 98, 102
Finance, 25, 35, 81, 88

Financial analysis, 36, 136
Financial capacity, 166, 175
Financial consortium, 34
Financial planning, 35
Financial sustainability, 110
Financing, 14, 27, 35, 37, 58, 59, 67, 85, 114, 116, 174, 176–178, 189, 191
Fishery resources, 171
Fishing, 2, 5, 8, 10, 12, 13, 55, 65, 164
Flora, 1, 21, 24
Floreana, 2–4, 9, 15, 49, 51–54, 166
Floreana Island, 51, 52
Food and Agricultural Organization of the United Nations (FAO), 52, 53
Food security, 52
Food supply, 22, 50, 103
Fossil fuels, 2, 3, 7, 11–17, 22, 24, 26, 27, 50, 52, 57, 59, 60, 63, 64, 68, 77, 79, 85, 87, 97, 153, 156, 158, 159, 165, 169, 186, 188, 189
Fragile ecosystems, 16, 17, 107, 108, 132, 135, 137
Fragile environments, 1–17, 21–30, 45, 54, 141, 159, 163–178, 183–193
Fragmentation, 3
Framework legislation, 169
Freshwater, 28
Frontier, 4–6
Fuel subsidies, 13, 186, 188
Full-cost accounting, 27
Functional unit, 95

G
Galapagos, 1–17, 21–26, 28, 30, 33–46, 66, 158, 160, 166, 186–192
Galapagos Explorer, 11, 50
Galapagos Islands, 1, 10, 21, 23–25, 29, 30, 45, 46, 49–55, 57, 66, 76, 93–104, 118, 141–151, 156–157, 189
Galapagos Islands Zero Fuel Fossil Initiative, 14
Galapagos Marine Reserve, 21
Galapagos National Park (GNP), 5, 188
Galapagos tortoises, 2
Garbage, 23, 153
Generating capacity, 46, 58, 74, 81, 85, 86
Generators, 5, 7, 12, 15, 21, 28, 49, 52–54, 65, 71, 73–75, 79, 80, 121, 141
Geographic, 85, 164, 166, 168, 176, 185, 190, 192, 193
Geographic isolation, 165
Geographic positioning system (GPS), 36
Geopolitical, 167, 185
Geopolitics, 163, 174, 190

Geothermal, 29, 57–63, 67, 69, 72, 76–80, 85, 86, 169, 178
Geothermal energy, 34, 58, 60–63, 69, 86
Germany, 14, 51, 68–75, 79, 84, 86, 87, 114, 138, 175
GHG, *see* Greenhouse gas (GHG)
GHG emissions, *see* Greenhouse gas (GHG) emissions
Glacial meltwater, 33
Global Environment Facility (GEF), 24
Global Environment Fund (GEF), 25
Global horizontal irradiance, 36
Glycerol, 52
Good living "sumak kawsay," 108
Gran Canaria, 28
Green electricity, 74
Green Energy Act (GEA), 82, 84
Green Energy Act (Ontario), 168
Greenhouse gas (GHG), 41, 93, 97, 100–102, 159, 160
Greenhouse gas (GHG) emissions, 29, 63, 94, 96–98, 100–102, 104, 118, 137, 151, 155, 169, 171
Grid access, 68, 70, 81, 86, 88
Grid connection, 37, 75, 78, 81, 88
Grid extension, 108
Grid infrastructure, 37, 60
Grid mix, 34
Grid-supply tariff (GST), 59, 60
Gross National Product (GNP), 188
Groundwater, 58, 62
Growth, 6–8, 10, 12, 17, 38, 44, 45, 50, 55, 72, 74, 76, 79, 98, 102, 142, 159, 168, 170, 171, 185
Growth rate, 8, 49, 169
Guyana, 22, 107

H
Habitats, 3, 23, 50, 55, 83, 132, 156, 164
Haida Gwaii, 63–67, 166
Haiti, 22
Hawaiian Islands, 59, 60
Health, 4, 29, 34, 63, 83, 113, 132–134, 136–138, 155, 156, 159, 168, 170
Heating fuel, 29
Holistic approach, 157
Homeowners, 59, 60, 109
Hostels, 39–41
Hotels, 7, 39–46, 190
Hot water systems, 59, 85
Households, 8, 25, 34, 60, 77, 78, 108–110, 113, 124, 130, 136, 154, 155, 157, 188
Household survey, 124

Human capital, 159
Human Development Index (HDI), 142
Human-environment interactions, 136
Human health, 168
Hurricanes, 59, 85
Hybrid power systems, 175
Hydrocarbons, 2, 5, 11, 13, 24, 165, 167, 174, 177
Hydroelectric energy, 169
Hydrogen gas, 65
Hydropower, 69, 72, 76–78, 80, 82
Hysteresis, 26

I
Iceland, 60–63, 85, 86
Iconic species, 2, 16, 22
Iguanas, 2, 11, 50, 93
Immigration, 23
Incentives, 13, 27, 29, 34, 38, 57, 61, 62, 69, 75, 83, 86, 170, 186, 189
Indian Act 1876 (Canada), 173
Indian Ocean, 58
Indicators, 36, 108, 112, 132
Indigenous, 66, 108–110, 113, 119, 136, 137, 164, 185, 190
Indigenous peoples, 66, 113, 136
Indonesia, 60–63, 69
Industrial solid waste, 134, 135
Information gaps, 175–177
Infrastructure, 9, 22, 26, 29, 37, 38, 58, 60, 70, 97, 102, 114, 118, 141, 155, 164, 167–169, 171, 174, 176, 185, 187, 188, 190, 193
Innovation, 10, 26, 27, 29, 46, 59, 67, 86, 87, 158, 178, 187, 188
Insects, 3, 16, 132, 133
Installation, 37, 51, 59, 60, 70–73, 77, 109–111, 113, 115, 125, 128–133, 135, 156–158, 190, 191, 224
Installation process, 109, 129
Installed capacity, 60, 74, 76
Institutional barriers, 80
Integrated solid waste management, 134–135
Intensive energy, 93
Inter-American Development Bank, 27, 35, 36
Inter-American Institute for Cooperation on Agriculture (IICA), 51, 52
Interdisciplinary, 178, 191, 192
Interest rate, 81, 122
Intergenerational connections, 176, 188
Intermediate fuel oil (IFO), 11
International banks, 35
International Energy Agency (IEA), 42, 78, 79

International Finance Corporation (IFC), 36
International Galapagos Tour Operators Association (IGTOA), 25
Interrelationships, 184, 193
Interviews, 39, 42, 109, 119, 122, 129
Inuit, 164, 165, 173, 174
Invasive species, 2, 4, 11, 52
Invertebrates, 2–4, 11
Investment, 25, 27, 28, 35, 37–39, 61, 84, 86, 108, 110, 117, 122, 127, 165, 170, 171, 176, 177, 187, 188
Isabela Island, 4, 9, 10, 15, 49, 53, 54, 141–150
Isabella, 10, 49, 53
Island environments, 57–88
Island of Santa Cruz, 39, 45
Islands, 1–17, 21–30, 39, 41, 44–46, 49–55, 57–88, 93–104, 118, 141–151, 156–159, 164–166, 169, 174, 185, 186, 189, 190
Isle of Eigg, 166
Isolated, 30, 59, 76, 108, 109, 114, 123, 134, 138, 141, 142, 144, 151, 190
Isolated islands, 59, 141, 142
Isolation, 1–3, 16, 21, 22, 26, 118, 136, 165

J
Japan, 5, 14, 15, 57–59, 68, 77–79, 84–86, 88
Jatropha, 15, 51–55, 96, 98–99, 102, 103
Jatropha curcas, 15, 51–53, 98
Jessica, 11, 21, 50, 93
Jiudao Yakou village, 155–159
Jurisdictional authority, 172
Jurisdictions, 61, 63, 68, 69, 75, 77, 81, 84, 87, 88, 170–173, 175

K
Kamchatka Peninsula, 60
Key drivers, 136
Kiribati, 59
Knowledge transfer, 34
Kuril Islands, 60

L
Land-based tourism, 7, 39, 44, 49, 50, 55
Land claims settlement, 173
Landfill, 141–151, 157, 160
Landfill gas, 64, 82, 84
Landfill stabilization, 145–150
LandGEM model, 144

Land use, 103, 191
Land use planning, 171
Latin America, 36, 54, 81
Laws, 8, 25, 37, 57–88, 134, 170
LEED Gold sustainability certification, 189
Legal documentation, 37
Legal system, 68
Legislation, 62, 69, 70, 87, 134, 168, 169, 171, 173
Legitimacy, 117
Leopold matrix, 131
Lessons learned, 27, 29, 38–39, 46, 70, 76, 166, 175–177
Life cycle assessment (LCA), 93–104, 137
Life cycle inventory (LCI), 94, 95, 100
Life cycle stages, 95, 96, 104
Light bulbs, 15, 41, 42
Liquefied natural gas (LNG), 68, 97, 98, 102, 167, 173
Liquid biofuels, 54
Liquid fuels, 97, 98
LNG, *see* Liquefied natural gas (LNG)
Load management, 29, 70
Local knowledge, 158, 176
Logistics, 35, 38, 109, 166, 175
Longevity, 38
Low carbon, 26, 30, 33, 169
Low carbon scenario, 42
Low-cost, 44, 46, 154–159
Low-efficient energy, 154
Low-income households, 154

M
Manabi, 15, 51, 52, 98
Marine
 diesel, 55
 ecosystem, 93, 103
 energy, 50
 environment, 50, 55, 186
 iguana, 3, 11, 22, 50
 mammals, 164
 resources, 5, 141
 spills, 50, 94, 96, 103, 186
 tanker, 96
 transportation, 28, 93, 94
 vessels, 21, 26
Market
 assessment, 35, 36, 72
 based policies, 79
 pool price, 75
 value, 52
Matrix, 131, 132
Maximum cap, 70

Methane
    decay rate, 144–146
    generation, 141–151
    recovery rate, 141, 144
Methanol, 52
Methyl ester oil, 52
Micro-finance, 35
Migrants, 5, 8
Mineral rights, 61, 86
Mining, 63, 107, 164, 167, 173, 174, 177
Ministry of Electricity and Renewable Energy (MEER), 15, 50–52, 76, 109
Mitigation, 33, 133–135, 166, 167, 172, 174
Mobility, 109, 118
Model, 23, 64, 72, 76, 95, 100, 111, 113–118, 121, 123–127, 132, 136–138, 144, 146–150
Modelling, 23, 30
Modularity, 109
Monitor, 137, 190
Monitoring, 34, 82, 133, 134, 136
Moratorium, 9
Morocco, 28
Multilateral, 191
Municipal, 44, 83

N

National parks, 5, 6, 21, 66, 171, 185
National Sustainable Energy Policy of Barbados, 27
Native forest, 107
Natural capital, 159
Natural gas, 78, 96–98, 102, 103, 164, 165, 167, 169, 170, 173
Natural habitats, 132
Natural resources, 4, 6, 22, 23, 64, 97, 98, 158, 163–165, 167, 170, 171, 173, 174
Natural Resources Canada (NRCan), 28, 64, 65, 97, 98, 165, 170, 171, 174
Natural ventilation, 42
Nature reserves, 185
Net exporter, 96
Net importer, 96
Net power demand, 65
Net present value (NPV), 122
Networks, 22, 23, 29, 113–115, 134, 155, 158, 165, 176, 188, 191
Nevis, 22
Newfoundland, 64, 65, 169, 172
New towns, 168
New Zealand, 58–60
Nitrogen, 154

Nonindigenous, 107
Northern, 28–29, 34, 66, 70, 75, 79, 80, 163–178, 192
Northern Canada, 28–29, 171, 192
Northern energy development, 172, 173
Northwest Territories, 164, 171, 173
Nova Scotia, 67, 169
NO_x, 28
Numeric assumptions, 96
Nunavut, 164, 171, 173

O

Ocean, 1–3, 22, 50, 55, 58, 62, 64, 67, 163, 169
Ocean currents, 1
Off-grid, 165, 174
Offset, 26, 34, 41, 42, 63, 65, 102, 133
Offshore, 58, 64, 66, 69, 71–74, 79, 164, 168, 171
    drilling, 168
    oil drilling, 164
    wind farm, 66, 79
Oil
    press, 53
    spill, 2, 11, 16, 22, 93
Ontario, 29, 38, 82–85, 168
Operational capacity, 80, 189
Operational cost, 35, 71
Organic farming, 35, 36
Organic wastes, 143, 144, 146, 151
Outboard motors, 118–120, 122
Ownership rights, 61, 86
Oxidation, 53

P

Pacific islands, 59–60, 142, 143
Pacific Ocean, 1, 58, 62
Paradigms, 108, 115
Parque Nacionale Galápagos, 141
Partnerships, 24, 33, 35, 37, 51, 183, 186, 188, 191, 193
Path dependency, 26
Pathogens, 153
Pathways, 94–100, 102–104
Peak times, 75
Per capita, 142
Performance
    bonds, 35
    standards, 36
Periodic rate amendments, 69
Permafrost, 176
Phosphorous, 154

Photovoltaic, 15, 29, 30, 34, 50, 54, 59, 82, 108–111, 116, 120, 121, 123, 133, 134, 178
   generator, 15, 121
   solar home systems (SHS), 108
Pikaia Lodge, 25
Pipelines, 167, 168
Plant-based biofuels, 50
Polar bears, 164
Policy
   innovation, 46
   tools, 169
Political will, 25, 109, 191
Population growth, 8, 44, 45, 50, 142, 170, 185
Potable water, 43
Poverty abatement, 153
Power plants, 11, 22, 29, 59, 61, 69, 71, 73, 86, 98, 165, 178
Power purchase agreement (PPA), 25, 66–68, 76, 81, 87, 88
Precedents, 23, 27–30
Predators, 3, 103
Pre-feasibility, 36
Premium rates, 68, 78, 87
Price, 13, 14, 26, 41, 42, 62, 68, 69, 72–75, 77, 78, 81, 83, 87, 88, 165–168, 183, 186, 189
Price-adder mechanism, 83
Prince Edward Island, 64, 65, 86
Principles, 24, 27, 34, 68–70, 75, 170, 184
Problem-solving, 187
Procurement, 35, 37–38, 83, 84
Production
   bonus, 63
   standards, 52
Project
   development, 35–39, 82, 177
   financing, 177, 191
Protocols, 109, 135, 137, 176
Prototype, 113, 118, 138
Public education, 24, 64, 178
Public lands, 171
Public-private partnerships, 24
Public utilities, 59, 69, 81, 88
Puerto Ayora, 14, 15, 39, 41
Puerto Baquerizo, 10, 11, 50, 157
Purchase contracts, 37

Q
Qualitative analysis, 109
Quality of life, 108, 154, 173, 185
Quechua language, 108

R
Radical energy, 34–39
Rainfall variability, 33
Rainforest Alliance, 43
Rainwater capture, 156
Ramea Island, 64, 65, 86, 166
Real costs, 26
Recyclable waste, 141, 142, 149, 157
Recycling scenarios, 149
Refinery capacity, 96
Regulator, 23, 26, 36, 59–62, 69, 70, 76, 86, 87, 111, 117, 123–126, 128, 168, 170, 174, 175, 178
Regulatory regime, 62, 178
Relationship building, 29
Reliability-centered maintenance (RCM), 123–127, 129
Remote
   communities, 28, 165–167, 169, 175, 177
   environments, 165, 166, 176
   islands, 22–23, 57, 59–61, 63, 65, 67, 68, 85–87
   island states, 22–23
   location, 166, 177, 185
Remoteness, 26
Renewable energy, 11, 14–16, 24–29, 34, 38, 46, 50–54, 57, 58, 63, 64, 66, 68–70, 72–74, 76–82, 86, 87, 94, 107–138, 155, 158, 165, 166, 168–170, 174–178, 183, 186, 189–191, 193
Renewable energy sources, 26, 50, 53, 54, 70, 78–80, 158, 169–170, 183, 189
Renewable Energy Sources Act, 70
Renewable energy technologies, 25, 66, 72, 165, 168, 176, 178, 190
Renewable Energy Unit (UER), 114, 116
Renewable portfolio standards (RPS), 68, 77–87
Residential heating demand, 167
Resilience, 167
Resource extraction, 185
Restoration, 22
Retrofitting, 26, 27
RETScreen, 34, 36
Reverse osmosis, 45
Risk mitigation, 166
Risk perception, 166, 167, 176, 177
Risk Priority Index (RPI), 125
River transport, 118, 123
Roads, 22, 34, 35, 118, 167
Run-of-the-river, 29, 66
Rural and remote, 165, 166, 176
Rural communities, 108, 155
Rural cooperatives, 51

Index

Rural economic development, 51
Rural electricity, 108
Rural electrification, 108, 112, 113, 115–118, 126, 135–137
Rural population, 108

S

St. Kitts, 22
Samso island, 63–67, 74, 86, 87
San Cristobal, 4, 9–11, 14, 21, 24, 49, 50, 141, 144–151, 156–157, 159, 160, 186, 190
San Cristobal Island, 11, 14, 21, 24, 186
Sanitary landfill, 157
Santa Cruz, 9, 10, 12, 14, 15, 25, 39, 41, 44, 45, 49, 141–151, 156
Scale-free networks, 176
Scales, 4, 15, 26, 29, 30, 34, 39, 58, 73, 77, 83, 97, 113, 141, 154, 167, 168, 170, 174, 176–178, 183, 184, 189, 193
Scale-up, 176
Scaling up, 108
Scenarios, 9, 10, 30, 42, 97, 102, 137, 141, 144–150, 164, 172
Schematics, 37
Scholl-Canyon model, 144, 146, 147, 150
Science, 23, 36, 39, 46, 58, 184, 190, 191
Scope definition, 94–95
Scoping, 36
Scotland, 67, 166
Scuba diving, 49
Seabirds, 11, 164
Sea conditions, 22, 120
Sea ice loss, 164
Sea ice observations, 164
Sea level rise, 22, 93
Sea lions, 2–4, 11
Sea turtle, 22
Self-government, 173
Self-supply tariff (SST), 59
Seminomadic, 116
Settlement, 4, 115, 163, 164, 168, 173
Settlement pattern, 164, 168
Seymour Airport, 189
Shared values, 46, 184
Shareholders, 67
Shipping, 50, 164, 171
Shipping impacts, 164
Shuar people, 116
Site investigation, 35, 36
Site selection, 36
Small island developing states (SIDS), 58, 85, 165
Small islands, 58, 59, 77, 85, 165

Snorkelling, 49
Snow and Grimwood Report, 6
SO_2, 28
Social benefits, 34, 49, 113, 136
Social capital, 55, 188
Social ecological, 23, 29, 30, 103, 159, 184–186, 191–193
Social ecological systems, 23, 29, 30, 159, 184–186
Social economic benefits, 188
Social equity, 174
Social factors, 29, 174
Social learning, 38, 46, 158, 183, 188
Social license, 83, 167, 176
Social networks, 23, 188
Soil quality, 131
Solar
  canoes, 119–123
  energy, 14–16, 24, 29, 34–35, 38, 46, 57, 59–60, 73, 76, 79, 85, 119, 134, 168
  irradiation, 36
  microgrid, 29
  panels, 15, 34, 59, 85, 121, 124, 128, 168, 189
  photovoltaic (PV), 15, 29, 34, 37, 59, 69, 71–74, 76–79, 82, 83, 85, 109, 111, 123, 168, 178
  photovoltaic systems, 109
  PV panels, 59, 78, 85
  Quest, 190
Solid waste, 130–135, 141–151, 157
  management, 130, 134–135, 141–151
South America, 76
South China Sea, 58
Space heating, 60
Spain, 14, 34, 38, 84
Spatial distribution, 136, 176
Spot price, 62, 75
Stakeholder
  engagement, 38, 39, 136
  relations, 35
Standard Offer Program (RESOP), 82
Standard of living, 4, 8–10, 16, 188
Storage technology, 123
Strategic alliance, 114
Strategic bridging, 115
Subsea transmission, 66, 75
Subsea turbines, 169
Subsea volcanoes, 62
Subsidies, 12–14, 25, 61, 64, 80, 113, 186–189
Subsidization, 26
Subsidized diesel systems, 166
Subsidized electrical tariff, 123

Sulphur, 52, 53
Sulphur dioxide (SO_2), 28, 53
Supply chain, 97, 102
Supply security, 189
Sustainability
    assessment, 109, 111–113, 135, 137
    assessment model, 137
    labeling, 42
Sustainable
    development, 108, 133, 155, 185–186
    development strategy, 185–186
    energy, 1–17, 21–30, 33–46, 50–52, 54, 55, 61, 94, 95, 100, 103, 104, 109, 112, 137, 153–160, 163–180, 183–193
    energy mix, 1–17, 21–30, 43, 45, 46, 50–52, 54, 55, 94, 100, 103, 104, 163–178, 183–193
    energy policy, 27, 187
    technology, 155
    tourism, 28, 118
System
    boundaries, 95
    complexity, 184

T

Tariff
    adjustments, 72
    structures, 73, 175
Technical
    capacity, 24, 120, 185
    design, 116
    expertise, 174, 178
Technology
    mix, 79
    transfer, 156–159, 165, 170, 176, 186, 189, 190
Telecommunication, 138
Temperature, 1, 22, 42, 52, 61, 164, 165, 168, 174, 176
Temporal, 183
Terrestrial, 3, 24, 26, 50, 54, 55, 103, 167, 174, 176
Territorial government, 165, 169, 171, 175
Test site, 64
Thermal
    electricity, 51
    energy, 21
    mass, 42
Tidal energy, 67, 86, 169
Tidewater
    ports, 168, 173
    terminals, 167
Timber extraction, 107
Tokelau Island, 59
Tourism, 2, 5–8, 10–14, 16, 22–26, 28, 30, 39, 41, 44, 45, 49, 50, 55, 65, 66, 93, 107, 118, 141, 143, 157, 164, 185, 190
Tourist boat, 50, 55
Toxicity, 95
Trade agreements, 82–85, 88, 189
Trade offs, 95, 174, 186
Traditional Aboriginal lands, 167
Traditional knowledge, 136
Traditional lands, 173
Transboundary, 191
Transdisciplinary, 183–193
Transformation, 6, 10, 16, 52, 96, 98, 178, 193
Transition, 23–26, 28–30, 35, 38, 41, 43, 46, 51, 74, 137, 166, 169, 170, 173–175, 177, 186, 188–191
Transmission bottleneck, 75
Transmission infrastructure, 70
Transparency, 38, 63
Transportation, 7, 8, 10–14, 16, 17, 21–24, 26–29, 49, 58, 63, 68, 85, 93–98, 100–102, 109, 110, 113, 115, 118, 119, 131, 132, 137, 163–168, 173, 174, 185, 188, 190
Transportation costs, 58, 68, 85, 102, 113
Treaties, 171, 173
Triglycerides, 52
Triple bottom line, 111
Turbines, 14, 15, 28, 64–66, 71–74, 79, 87, 168, 169, 178, 189
Typology, 186, 190

U

Uncertainty, 23, 30, 38, 96, 164, 166, 167, 184, 185, 190
UNESCO World Heritage Site, 21, 66
UN Framework Convention on Climate Change (UNFCCC), 24
United Nations Development Programme (UNDP), 24
United Nations Education, Scientific, and Cultural Organization (UNESCO), 6, 21, 22, 28, 58, 66
Universidad San Francisco de Quito, 36
University of Calgary, 36, 39, 46, 158, 159
Upstream emissions, 97, 98, 100, 102
Urban centers, 107
Urbanization, 107, 136, 155
US Environmental Protection Agency, 97, 98, 100, 144
Utility ratepayers, 59
Utilization license, 61–63

V

Vector disease, 109
Vertrauensschutz (legitimate expectations), 69
Villages, 59, 66, 155–159
Volcanic rocks, 60, 86
Volcanoes, 1, 62

W

Waste heat, 66
Waste management, 64, 113, 123, 129, 130, 134, 135, 138, 141–151, 156–159
 system, 113, 123, 129, 138, 157, 159
Water
 audit, 39, 44, 46
 contamination, 63, 86
 inputs, 95
 supply, 44
 use, 43, 44, 103
Water-energy nexus, 43–45
Weather, 22, 156, 166, 168, 176
Wicked problems, 184, 185, 190
Wildlife disturbance, 30
Wind farm, 65, 66, 70, 71, 79, 83, 168
Wind-hydro system, 28
Wind power generation, 14, 65, 74, 87
Wind turbines, 14, 15, 28, 64, 65, 168, 189
Women, 36, 155
Workforce, 23, 173
World Energy Council, 174, 193
World Heritage sites, 6, 21, 50, 66, 156, 185
World Summit on Sustainable Energy in Fragile Environments, 186
World view, 167
World Wildlife Fund (WWF), 23, 41

Y

Yantsa Ii Etsari, 108–117, 119, 123–126, 129, 133, 134
Yukon, 164, 169
Yunnan province, 155, 158

Z

Zero-order methods, 144
Zoning, 37

CPSIA information can be obtained
at www.ICGtesting.com
Printed in the USA
LVHW02*2252140318
569849LV00002B/36/P